Program Theory-Driven Evaluation Science

Strategies and Applications

Program
Theory-Driven
Evaluation
Science

Strategies and Applications

Stewart I. Donaldson

Claremont Graduate University

Psychology Press
Taylor & Francis Group

New York   London

Psychology Press   Psychology Press
Taylor & Francis Group   Taylor & Francis Group
270 Madison Avenue   27 Church Road
New York, NY 10016   Hove, East Sussex BN3 2FA

© 2007 by Taylor & Francis Group, LLC
Psychology Press is an imprint of Taylor & Francis Group
Originally published by Lawrence Erlbaum and Associates

Printed in the United States of America on acid-free paper
10 9 8 7 6 5 4 3
International Standard Book Number-13: 978-0-8058-4671-3 (Softcover) 978-0-8058-4670-6 (Hardcover)

Library of Congress Cataloging-in-Publication Data

Catalog record is available from the Library of Congress

Visit the Taylor & Francis Web site at
http://www.taylorandfrancis.com

and the Psychology Press Web site at
http://www.psypress.com

To compassionate scientists working to improve the human condition

And to my wellspring of love, life, and endless admiration—

Michelle Anne, Candice Danae, Scott Ian, and Russell Ian Donaldson

Contents

Preface

The image on the cover hints that this book is about the use of a variety of tools. The tools of 21st-century evaluation science hold great promise for preventing and solving the social, community, and organizational problems likely to confront modern and developing societies in the years ahead.

This book stems out of my desire to help fill the gap between what we espouse in the 21st-century evaluation literature, and what appears to be happening on the ground in contemporary evaluation practice. Students and colleagues alike often yearn for published examples of how evaluation science actually unfolds in practice. This volume attempts to contribute to a broader effort by members of the evaluation community in recent times to fill this void. It provides detailed examples of how program theory-driven evaluation science has been used to develop people, programs, and organizations.

This is not another standard evaluation methods textbook. It is not intended to be another prescriptive text instructing how best to practice evaluation. There are many of those available today, some of them with quite different (sometimes opposing) views.

This is a book that outlines simple and commonly accepted strategies for practicing evaluation, followed by detailed accounts of how those strategies have actually played out in the face of the complexities and challenges of "real world" evaluation settings. The "real" evaluations presented here are not put forth as "exemplars" or "outstanding evaluations." Instead the goal of this volume is to illustrate the authentic challenges of implementing an evaluation approach in practice.

Some might say, or at least think, it is very bold and risky to disclose one's dirty laundry about the trails and tribulations of working with stakeholders, and within the limiting constraints often imposed by the realities of contemporary evaluation practice. However, I've decided it is worth the risk of criticism, for a chance at improving future efforts to use evaluation science to prevent human suffering and to promote human welfare and achievement.

In the chapters ahead, I attempt to provide a "state-of-the-art" treatment of the practice of program theory-driven evaluation science. It is my goal to fill a void in the extant literature, namely a lack of detailed examples of program theory-driven evaluation science being implemented in "real world" settings. An in-depth description of the nuances and results from a series of "authentic program theory-driven evaluations" from recent evaluation practice are presented. The presentation of each case will be organized around a concise three-step model for conducting program theory-driven evaluation science:

1. Developing program impact theory.
2. Formulating and prioritizing evaluation questions.
3. Answering evaluation questions.

In chapters 1 through 3, the emergence of program theory-driven evaluation science and specific strategies for carrying out each of the three steps above are presented. After clearly defining and describing program theory-driven evaluation science, recent applications are presented in chapters 4 through 8 to illustrate the process and potential products of this evaluation approach. These applications include evaluations of four programs within a large, statewide Work and Health Initiative——the Winning New Jobs Program, the Computers in Our Future Program, the Health Insurance Policy Program, and the Future of Work and Health Program. Chapters 10 and 11 discuss a fifth case and provide an example of an evaluation proposal based on a program theory-driven evaluation science framework.

Challenges and lessons learned from the cases presented, and from the program theory-driven evaluation literature more broadly, are described in chapter 12. This discussion includes crosscutting lessons learned from the Work and Health Initiative, as well as from a range of other program domains. Finally, practical implications of the emerging transdiscipline of evaluation science are discussed in chapter 13.

This book is likely to be of great value to practicing evaluators, evaluation scholars, teachers, and trainers of evaluation and applied research, advanced undergraduate and graduate students, and other professionals interested in learning how evaluation science can be used in complex "real world" settings to improve people, programs, and organizations. It can be used as a text or supplementary reader for advanced undergraduate and graduate courses in evaluation and applied research methods. This book should also be of interest to those studying topics in work and health, career development, human service organizations, and organizational improvement and effectiveness.

ACKNOWLEDGMENTS

Many colleagues and graduate students worked with me over the years on the Work and Health Initiative evaluation. Although the list of contributors is much too long to display here, a heartfelt thanks to all of you who were part of our highly productive evaluation team.

The Work and Health Initiative evaluation was supported by a grant from The California Wellness Foundation to Dr. Stewart I. Donaldson, Principal Investigator. Special thanks to Gary Yates, Tom David, Ruth Brousseau, and Lucia Corral Pena of The California Wellness Foundation; Sharon Rowser and Rick Price of the Wining New Jobs Program; Helen H. Schauffler and E. Richard Brown of the Health Insurance Policy Program; Linda Fowells, Paul Vandeventer, Wendy Lazarus, and George Gundry of the Computers In Our Future Program; Ed Yelin of the Future of Work and Health Program; and to the leadership and staff of the many grantee organizations for their endless commitment to improving the health and well-being of California workers and their families.

No one deserves more recognition and appreciation than Dr. Laura Gooler. Dr. Gooler successfully directed the Work and Health Initiative evaluation for more than 5 years. She coauthored most of the evaluation reports for the project, and has collaborated with me on a number of articles describing findings and lessons learned. Her excellent contributions provided the foundation for most of the application chapters in this book. Thanks also to Marvin Alkin, Christina Christie, Nicole Eisenberg, and Lynn Winters for allowing me to include my contribution to their excellent project as another application of program theory-driven evaluation science.

My editorial assistant, Shanelle Boyle, has been a gem to work with on this project. I owe a sincere debt of gratitude for her commitment to excellence, and her positive outlook as we spent many tedious hours editing and finalizing the production of this manuscript.

I am grateful for the helpful reviews and suggestions for improvement provided by reviewers Thomas Chapel, Centers for Disease Control, David Fetterman, Stanford University, Melvin Mark, The Pennsylvania State University, and Donna Mertens, Gallaudet University. Special thanks to Debra Riegert and Rebecca Larsen of Lawrence Erlbaum Associates, Inc., for helping me address reviewer concerns, as well as for providing many useful suggestions themselves.

It is only with the help of all these great friends, students, and colleagues that I was able to complete this book. In the end, it will be well worth the effort if it helps you and others better understand the benefits and challenges of using this powerful tool called program theory-driven evaluation science to better the human condition.

—*Stewart I. Donaldson*

Program Theory-Driven Evaluation Science

Strategies and Applications

FOUNDATIONS AND STRATEGIES FOR PROGRAM THEORY-DRIVEN EVALUATION SCIENCE

1

The Emergence of Program Theory-Driven Evaluation Science

This book is about using evaluation science to help promote human welfare and improve the quality of human lives in the 21st century. It focuses on ways to use a modern tool, generally referred to as *evaluation science*, to develop people, as well as groups of people working toward common goals. These social groups are commonly known as *organizations*.

Improving programs and organizations designed to prevent and/or ameliorate social, health, educational, community, and organizational problems that threaten the well-being of people from all segments of our global community is the pursuit at hand. More specifically, this book explores applications of how evaluation science can be used to develop and improve programs and organizations dedicated to promoting health, well-being, human productivity, and achievement. Given that we live in a time of unprecedented change, uncertainty, and human suffering, this book will indeed be exploring the application and potential of a very useful, powerful, and critically important human-designed tool.

The history of humankind is filled with examples of the development and refinement of tools created to solve the pressing problems of the times (e.g., various types of spears in the hunter–gatherer days, machines in the industrial era, and computers in the information age). The modern day tool I refer to as evaluation science has evolved over the past three decades, and is now being widely used in efforts to help prevent and ameliorate a variety of human and social problems.

As with other tools, many creative minds have tried to improve on earlier prototypes of evaluation. These vigorous efforts to improve the effectiveness of evaluation, often in response to inefficiencies or problems encountered in previous attempts to promote human welfare, have left us with a range of options or brands from which to choose (see Donaldson & Scriven, 2003a). For example, now available on the shelf at your global evaluation tool store are:

- Traditional social science brands (Shadish, Cook, & Campbell, 2002).
- Transdisciplinary brands (Scriven, 2003).
- Empowerment brands (Fetterman, 2003).
- Results-oriented management brands (Wholey, 2003).
- Fourth generation brands (Lincoln, 2003).
- Inclusive brands (Mertens, 2003).
- Feminist brands (Seigart & Brisolara, 2002).
- Utilization-focused brands (Patton, 1997).
- Realists brands (Mark, Henry, & Julnes, 2000; Pawson & Tilley, 1997).
- Theory-driven evaluation science brands (Chen, 1990; Donaldson, 2003; Rossi, Lipsey, & Freeman, 2004; Weiss, 1998) among others.

This proliferation of evaluation approaches has led evaluation scholars to spend some of their professional time developing catalogues profiling the tools and suggesting when we might use a particular brand (e.g., Mark, 2003; Christie, 2003). Others have taken a different approach and tried to determine which brands are superior (e.g., Shadish, Cook, & Leviton, 1991), while yet another group has strongly argued that some brands are ineffective and should be kept off the shelf altogether (e.g., Scriven, 1998; Stufflebeam, 2001).

One potential side effect of the rapid expansion of brands and efforts to categorize and critique the entire shelf at a somewhat abstract level is that a deep understanding of the nuances of any one evaluation approach is compromised. It is precisely this detailed understanding of the potential benefits and challenges of using each evaluation approach across various problems and settings that will most likely advance contemporary evaluation practice.

Therefore, the main purpose of this book is to examine arguably one of the most evolved and popular evaluation brands in detail—*Program Theory-Driven Evaluation Science*. This is accomplished by exploring findings and lessons learned from several completed program theory-driven evaluations. In addition, one chapter describes a proposal of how to use program theory-driven evaluation science to address a hypothetical evaluation problem presented and critiqued by scholars studying the differences between modern evaluation approaches (Alkin & Christie, 2005). These cases and examples illustrate how program theory-driven evaluation science can be used to develop and improve programs and organizations, and suggest ways to refine and simplify the practice of program theory-driven evaluation science.

COMPREHENSIVE THEORY OF EVALUATION PRACTICE

History of Evaluation Theory

Shadish et al. (1991) examined the history of theories of program evaluation practice and developed a three-stage model showing how evaluation practice evolved from an emphasis on truth, to use, and then toward integration. The primary assumption in this framework is that the fundamental purpose of program evaluation theory is to specify feasible practices that evaluators can use to construct knowledge of the value of social programs that can be used to ameliorate the social problems to which the programs are relevant.

The criteria of merit used to evaluate each theory of evaluation practice were clearly specified by Shadish et al. (1991). A good theory of evaluation practice was expected to have an excellent knowledge base corresponding to:

1. Knowledge—what methods to use to produce credible knowledge.
2. Use—how to use knowledge about social programs.
3. Valuing—how to construct value judgments.
4. Practice—how evaluators should practice in "real world" settings.
5. Social Programming—the nature of social programs and their role in social problem-solving.

In their final evaluation of these theories of evaluation practice, Shadish et al. (1991) concluded that the evaluation theories of Stage I theorists focused on truth (represented by Michael Scriven and Donald Campbell); Stage II theorists focused on use (represented by Joseph Wholey, Robert Stake, and Carol Weiss); and that both Stage I and Stage II evaluation theories were inadequate and incomplete in different ways. In other words, they were not judged to be excellent across the five criteria of merit just listed.

In contrast, Stage III theorists (Lee Cronbach and Peter Rossi) developed integrative evaluation theories that attempted to address all five criteria by integrating the lessons learned from the previous two stages. They tried not to leave out any legitimate practice or position from their theories of practice, and denied that all evaluators ought to be following the same evaluation procedures under all conditions. Stage III theories were described as *contingency theories of evaluation practice*, and were evaluated much more favorably than the theories of practice from the previous two stages.

In Shadish et al.'s (1991) groundbreaking work on theories of evaluation practice, Peter Rossi's seminal formulation of theory-driven evaluation was classified as a *Stage III Evaluation Theory of Practice*, the most advanced form of evaluation theory in this framework, and was evaluated favorably across the five criteria of merit. Theory-driven evaluation was described as a comprehensive attempt to resolve dilemmas and incorporate the lessons from the applications of past theories to evaluation practice; it attempted to incorporate the desirable features of past theories without distorting or denying the validity of these previous positions on how to practice program evaluation.

Shadish et al. (1991) noted that the theory-driven evaluation approach was a very ambitious attempt to bring coherence to a field in considerable turmoil and debate, and that the integration was more or less successful from topic to topic. Rossi's (2004) theory-driven approach to evaluation offered three fundamental concepts to facilitate integration:

1. Comprehensive Evaluation—studying the design and conceptualization of an intervention, its implementation, and its utility.
2. Tailored Evaluation—evaluation questions and research procedures depend on whether the program is an innovative intervention, a modification or expansion of an existing effort, or a well-established, stable activity.
3. Theory-Driven Evaluation—constructing models of how programs work, using the models to guide question formulation and data gathering; similar to what econometricians call model specification.

These three concepts remain fundamental to the discussion that follows in this book about how to use program theory-driven evaluation science to develop and improve modern programs and organizations.

The notion of using a conceptual framework or program theory grounded in relevant substantive knowledge to guide evaluation efforts seemed to take hold in the 1990s. The work of Chen and Rossi (1983, 1987) argued for a movement away from atheoretical, method-driven evaluations, and offered hope that program evaluation would be viewed as and become more of a rigorous and thoughtful scientific endeavor. Chen (1990) provided the first text on theory-driven evaluation, which became widely used and cited in the program evaluation literature. Furthermore, two of the most popular (best-selling) textbooks on program evaluation (Rossi, Lipsey, & Freeman, 2004; Weiss, 1998) are firmly based on the tenets of theory-driven evaluation science, and offer specific instruction on how to express, assess, and use program theory in evaluation.

Although Shadish et al.'s (1991) framework has been widely cited and used to organize the field for over a decade, it is now controversial in a number of ways. For example:

1. Were all the important theorists and theories of practice included?
2. Was each position accurately represented?
3. Was the evaluation of each theory of practice valid?

I seriously doubt that most of the theorists evaluated and their followers would answer "yes" to all three questions. On the contrary, I would suspect arguments that the authors' biases, assumptions, or agendas influenced how they described others' work and characterized the field of evaluation theory.

Looking Toward the Future

Although knowledge of the history of evaluation theories can provide us with important insights, practicing evaluators tend to be most concerned with current challenges and the future of evaluation practice. More than a decade after Shadish et al.'s (1991) work, Donaldson and Scriven (2003b) employed an alternative methodology to explore potential futures for evaluation practice. Instead of describing, classifying, or evaluating others' work over the past three decades, they invited a diverse group of evaluators and evaluation theorists to represent their own views at an interactive symposium on the future of evaluation practice. A "last lecture" format was used to encourage each participant to articulate a vision for "How We Should Practice Evaluation in the New Millennium." That is, based on what each evaluator had learned over her or his evaluation career, she or he was asked to give a last lecture, passing on advice and wisdom to the next generation about how we should evaluate social programs and problems in the 21st century. In addition to six vision presentations, five prominent evaluators were asked to give a reaction lecture about the diverse visions and future of evaluation practice. These reaction lectures were followed by responses from the initial visionaries and the audience of more than 300 participants.

This approach to understanding theories of evaluation practice led to the expression of some of the most diverse and thought-provoking ideas about how to best evaluate social programs and problems (Donaldson & Scriven, 2003b; Slater, 2006; Triolli, 2004). One striking aspect of this rich discourse was that program theory or program theory-driven evaluation seemed to be an integral component of many of the visions and modern theories of evaluation practice (Crano, 2003). That is, in addition to being a stand-alone comprehensive and contingency theory of evaluation practice, now supported by published examples (Donaldson, 2003; Weiss, 2004), many elements of theory-driven evaluation have been incorporated into other theories of evaluation practice. For example, (a) the transdisciplinary view of evaluation discusses the enlightened use of and the tripartite role of program theories (the alleged program theory, the real logic of the program, and the optimal program theory; Scriven, 2003); (b) empowerment evaluation describes how its central process creates a program theory (Fetterman, 2003); (c) inclusive evaluation encourages evaluators to prevent program theories from contributing to negative stereotyping of marginalized groups (Mertens, 2003); (d) results-oriented management incorporates logic models and program theory to guide performance measurement (Wholey, 2003); and (e) an integrative vision for evaluation theories suggests theory-driven evaluation has achieved enough successes to have earned a place in the toolkit of evaluation approaches (Mark, 2003). In other recent work, realist evaluation focuses on program mechanisms and incorporates moderators as part of program theory and evaluation (Mark et al., 2000; Pawson & Tilley, 1997). Utilization-focused evaluation is highly compatible with theory-driven evaluation (Christie, 2003) and often makes use of program theory in evaluation (Patton, 1997). Furthermore, advances in construct validity and social experimentation are making theory-driven evaluation a much more precise and promising endeavor (Crano, 2003; Shadish et al., 2002).

REFINEMENTS FOR CONTEMPORARY EVALUATION PRACTICE

Many of the prior writings on theory-driven evaluation have inspired evaluators to move away from method-driven evaluation approaches and to use program theory to improve programs and evaluations (e.g., Chen, 1990; Chen & Rossi, 1987). However, recent criticisms suggest that although there has been an explosion of literature on the topic in recent years, many of these writings are at a "stratospheric level of abstraction" (Weiss, 1997, p. 502). Much has been learned about the nuances of practicing theory-driven evaluation during the past decade, but the dissemination of practical examples and lessons learned appears to be needed to make this approach more straightforward and accessible to the evaluation, program design, and organizational development communities. Finally, there continues to be concern and misunderstandings about the value and feasibility of conducting theory-driven evaluations in modern settings (Donaldson, 2003). One purpose of this book is to explore some of these concerns and issues in the context of real examples from current evaluation practice, and to simplify and update this increasingly common theory of evaluation practice.

Defining Program Theory-Driven Evaluation Science

As evidenced from the literature, the use of program theory has become commonplace in evaluation practice, and has been diffused or incorporated into the most popular approaches or theories of evaluation practice. Gargani (2003) recently offered a rather detailed account of the history of the use of program theory in evaluation, and concluded that it has a very long history indeed. He claimed that the practice of articulating and testing program theory was first introduced to evaluation by Ralph W. Tyler in the 1930s, but did not find widespread favor with evaluators for more than 50 years after it was introduced. Gargani (2003) notes that the practice of articulating and testing program theory appears to be one of the few evaluation approaches that, although not universally endorsed by evaluation theorists, is widely applied by evaluation practitioners.

Donaldson and Lipsey (2006), Donaldson (2003), and Weiss (1997) noted that there is a great deal of confusion today about what is meant by theory-based or theory-driven evaluation, and the differences between using program theory and social science theory to guide evaluation efforts. For example, the newcomer to evaluation typically has a very difficult time sorting through a number of closely related or sometimes interchangeable terms such as *theory-oriented evaluation, theory-based evaluation, theory-driven evaluation, program theory evaluation, intervening mechanism evaluation, theoretically relevant evaluation research, program theory, program logic, logic modeling*, and the like. Rather than trying to sort out this confusion, or attempt to define all of these terms and develop a new nomenclature, a rather broad definition is offered in this book in an attempt to be inclusive.

> Program Theory-Driven Evaluation Science is the systematic use of substantive knowledge about the phenomena under investigation and scientific methods to improve, to produce knowledge and feedback about, and to determine the merit, worth, and significance of evaluands such as social, educational, health, community, and organizational programs.

Program theory-driven evaluation science is often used to (a) develop and improve programs and organizations focused on preventing and solving a wide range of pressing human concerns and problems, (b) to aid decision making, (c) to facilitate organizational learning and the development of new knowledge, and (d) to meet transparency and accountability needs.

The typical application involves using program theory to define and prioritize evaluation questions, and using scientific methods to answer those questions. Thus, I prefer to describe this evaluation approach as program theory-driven but empirically based evaluation, as opposed to theory-based evaluation. However, I recognize that others may be using the same or a very similar set of evaluation procedures but describing them with different terms.

It is important to note that because program theory-driven evaluation science attempts to incorporate best practices, procedures, and methods from other theories of evaluation practice, evaluations conducted under different labels may fit within the definition of program theory-driven evaluation science.

However, although there are many possible ways to practice or add variations to the practice of program theory-driven evaluation science, in an effort demystify this evaluation approach, I attempt to provide a very straightforward, basic "backbone" or foundation that can be applied across a wide range of modern program evaluation problems and settings.

Therefore, in an effort to incorporate various lessons learned from the practice of program theory-driven evaluation in recent years (e.g., Donaldson & Gooler, 2003), and to clarify and simplify this approach to make it more accessible to a wider range of evaluation practitioners, the following simple three-step model is proposed for understanding the basic activities of *Program Theory-Driven Evaluation Science*:

1. Developing program impact theory.
2. Formulating and prioritizing evaluation questions.
3. Answering evaluation questions.

Simply stated, evaluators typically work with stakeholders to develop a common understanding of how a program is presumed to solve a problem or problems of interest. Social science theory and prior research (if they exist) can be used to inform this discussion, and to assess the feasibility of the proposed relationships between a program and its desired initial, intermediate, and long-term outcomes (Donaldson & Lipsey, 2006). This common understanding or program theory helps evaluators and stakeholders identify and prioritize evaluation questions.

Evaluation questions of most interest are then answered using the most rigorous scientific methods possible given the practical constraints of the evaluation context.

The phrase *program theory-driven* (instead of theory-driven) is intended to clarify the meaning of the use of the word "theory" in this evaluation context. It aspires to clarify and specify the type of theory (e.g., program theory, not necessarily social science theory) that is guiding the evaluation questions and design. *Evaluation science* (instead of evaluation) is intended to underscore the use of rigorous scientific methods (i.e., qualitative, quantitative, and mixed-method designs) to attempt to answer valued evaluation questions. A renewed emphasis in evaluation practice on relying on systematic, scientific methods is especially important for overcoming the profession's negative reputation in some contexts (Donaldson, 2001b). That is, in some settings, evaluation is criticized for being an unreliable, soft, or a second-class type of investigation. The term *evaluation science* signals the emphasis placed on the guiding principle of systematic inquiry (Guiding Principles for Evaluators, 2004) and the critical evaluation standard of accuracy (Joint Committee on Standards for Educational Evaluation, 1994).

Program theory-driven evaluation science is essentially method neutral. That is, the focus on the development of program theory and evaluation questions frees evaluators initially from method constraints. The focus on program theory often creates a superordinate goal that helps evaluators get past old debates about which methods are superior to use in evaluation practice (Donaldson, 2003, 2004, 2005). From this contingency point of view, it is believed that qualitative, quantitative, and mixed-method designs are neither superior nor applicable in every evaluation context (cf. Chen, 1997). Whether an evaluator uses case studies, observational methods, structured or unstructured interviews, online or telephone survey research, a quasi-experiment, or a randomized controlled experimental trial to answer the key evaluation questions, is dependent on discussions with relevant stakeholders about what would constitute credible evidence in this context, and what is feasible given the practical and financial constraints.

One of the best examples to date of program theory-driven evaluation science in action is embodied in the Centers for Disease Control's (CDC) six-step Program Evaluation Framework (Figure 1.1). This framework is not only conceptually well developed and instructive for evaluation practitioners, but it also has been widely adopted for evaluating federally funded public health programs throughout the United States. The CDC framework extends the more concise, three-step program, theory-driven evaluation science model described, and offers additional details to help guide practitioners:

1. Engage stakeholders—those involved, those affected, and primary intended users.
2. Describe the program—needs, expected effects, activities, resources, stage, context, and logic model.
3. Focus the evaluation design—purpose, users, uses, questions, methods, and agreements.
4. Gather credible evidence—indicators, sources, quality, quantity, and logistics.
5. Justify conclusions—standards, analysis and/or synthesis, interpretation, judgment, and recommendations.
6. Ensure use and share lessons learned—design, preparation, feedback, follow-up, and dissemination.

The second element of this framework is a set of 30 standards for assessing the quality of the evaluation effort (adapted from the Joint Committee on Standards for Educational Evaluation, 2003). These standards are organized into the following four categories:

1. Utility—serve the information needs of the intended users.
2. Feasibility—be realistic, prudent, diplomatic, and frugal.
3. Propriety—behave legally, ethically, and with due regard for the welfare of those involved and those affected.
4. Accuracy—reveal and convey technically accurate information.

Finally, program theory-driven evaluations are also designed to strictly adhere to the American Evaluation Association's Guiding Principles (Guiding Principles for Evaluators, 2004):

1. Systematic Inquiry—evaluators conduct systematic, data-based inquiries.
2. Competence—evaluators provide competent performance to stakeholders.
3. Integrity and/or Honesty—evaluators display honesty and integrity in their own behavior, and attempt to ensure the honesty and integrity of the entire evaluation process.
4. Respect for People—evaluators respect the security, dignity and self-worth of respondents, program participants, clients, and other evaluation stakeholders.
5. Responsibilities for General and Public Welfare—evaluators articulate and take into account the diversity of general and public interests and values that may be related to the evaluation.

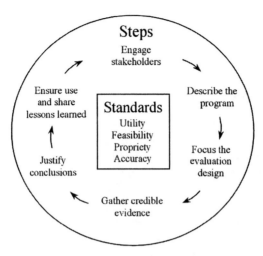

Figure 1.1.   Centers for Disease Control Evaluation Framework. From Centers for Disease Control and Prevention. *Framework for program evaluation in public health.* MMWR 1999; 48(No. RR–11). Available from http://www.cdc.gov/eval/framework. htm#graphic

Potential Benefits of Theory-Driven Evaluation Science

The hallmark of program theory-driven evaluation science is that it forces evaluators to try to understand the evaluand (the program) in some detail before rushing to action (e.g., employing their favorite methods to collect data). Program theory-driven evaluators are required to become familar with the substantive domains related to the problems of interest (or have that expertise on the evaluation team) in order to help formulate, prioritize, and answer key evaluation questions. This process or first step promises to address one of the fundamental challenges in evaluation practice today—Type II Errors or the false reporting of null results (see Lipsey, 1988; Lipsey & Wilson, 1993). For example, the wise use of program theory in evaluation can dramatically improve evaluation design sensitivity and validity by:

- Helping to disentangle the success or failure of program implementation ("action theory") from the validity of the program theory ("conceptual theory").
- Serving as a basis for informed choices about evaluation methods.
- Helping to identify pertinent variables and how, when (e.g., dose-response and program-decay functions), and on whom they should be measured.

- Facilitating the careful definition and operationalization of the independent (program) variables.
- Helping to identify and control for extraneous sources of variance.
- Alerting the program developer and evaluator to potentially important or intrusive interactions (e.g., differential participant response to the intervention).
- Dictating the proper analytical or statistical model for data analysis and the tenability of the assumptions required in that model.
- Helping to facilitate a thoughtful and probing analysis of the validity of program evaluations in a specific context and provide feedback that can be used to improve the program under investigation while developing a cumulative wisdom about how programs work and when they work (cf. Bickman, 1987; Chen, 1990; Donaldson, 2001a, 2003; Lipsey, 1993).

Furthermore, program theory-driven evaluation science leaves room for a variety of methodological approaches for producing credible knowledge, and often leads to immediately practical knowledge that can be used to improve programs and the organizations responsible for program success. As stated, it is one of the most promising evaluation approaches for producing cumulative knowledge and enlightening various stakeholders (including policymakers and project sponsors) about the problems of concern (Weiss, 1998).

Another important advantage of the three-step model for conducting the program theory-driven evaluation science approach presented previously is that it specifies feasible practices that can be used across a wide range of program settings. For example, many books on evaluation research seem to emphasize evaluation methods for efficacy evaluations. Efficacy evaluations address the question of whether a program or intervention can have effects under researchlike or "ideal conditions" (see Donaldson, 2003). Experimental and quasi-experimental designs are often touted as the preferred evaluation designs under these conditions (Lipsey & Cordray, 2000; Shadish et al., 2002). In contrast, many evaluations today are effectiveness evaluations and answer the question of whether a program has effects in "real world" settings (Fitzpatrick, 2002). Program theory-driven evaluation science is one of a few approaches that can shine in both types of evaluation contexts. For example, the tailoring of methods to the type of program and evaluation questions provides a very cost-effective approach for effectiveness evaluation. Designing programs and evaluations using prior theory and research resonates well with scholars conducting efficacy evaluations to advance knowledge in the substantive domains.

On a more practical level, the participatory approach to program theory-driven evaluation science that was used in the cases that will be described in this book has the additional potential advantage of engaging stakeholders in the evaluation process. Inclusion and collaboration during each step promises to enhance both the validity and utilization of the evaluation findings. I return and expand on this discussion of the potential benefits of program theory-driven evaluation science in chapter 12, after I have explored program theory-driven evaluation science in practice across a range of settings.

TOWARD A SCIENCE OF DEVELOPING PEOPLE AND ORGANIZATIONS

Human Transformation Technologies

Many social and organizational conditions hinder the healthy development and performance of a wide range of children, adolescents, working adults, and the elderly in contemporary societies (Donaldson & Scriven, 2003b). Well-intentioned practitioners, such as health and human service providers, educators, human resource and organizational development professionals and the like, design and implement programs and interventions to prevent dysfunction and to foster productivity and healthy human development. Hasenfeld (1983) provided a useful typology for classifying three general types of these "human transformation technologies":

1. People Processing Technologies: These attempt to transform clients by conferring on them a social label and a public status that evoke desirable reactions from other social units (e.g., a gifted child ready for an accelerated education program or cancer patient in need of special treatment).
2. People Sustaining Technologies: These attempt to prevent, maintain, and retard the deterioration of the personal welfare and well-being of clients (e.g., nursing home care for the elderly and income maintenance services for the poor).
3. People Changing Technologies: These aim directly at altering the personal attributes of clients in order to improve their well-being (e.g., education, counseling services, and personal and career development).

Program theory-driven evaluation science aspires to be a valuable partner in these noble efforts of human betterment (evaluation as a helping profession; Donaldson, 2001b). It is specifically concerned with developing productive collaborations with human service practitioners (of all sorts) in an effort to ensure program resources are deployed in the most effective and efficient ways possible. When done well, program theory-driven evaluation science accurately sorts out successes from failures, and provides useful scientific information for improving services designed to meet critical human needs.

Organizational Development

Most programs consisting of various human transformation technologies are delivered by organizations (including government and public service, nonprofit, and for-profit organizations), and sometimes by units within organizations (e.g., human resources or organizational effectiveness departments). *Organizational development* is commonly defined as a process that applies behavioral science knowledge and practices to help organizations achieve greater effectiveness (Cummings & Worley, 2005). This process often involves improving the well-being and productivity of organizational members, as well as improving organizational effectiveness.

Improving Organizational Members. Organization development programs focused on improving the quality of work life and the performance of organizational members typically fall into three general categories: *human process interventions, technostructural interventions*, and *human resource management interventions*. Human process interventions generally focus on improving working relationships and include t-groups, conflict resolution programs, and team building initiatives. Technostructural programs focus on modifying the technology and the structure of organizations and include interventions such as downsizing, reengineering, total quality management, job design, and job enrichment programs. Human resource management interventions deal with managing individual and group performance and typically include goal setting interventions, performance appraisal programs, reward systems, career planning and development interventions, employee wellness programs, work-family balance efforts, and workplace diversity programs.

Although it is true that most of these programs designed to improve employee productivity and well-being were initially designed and implemented in corporate contexts, in recent years they are being widely applied in health care, school systems, the public sector, global settings, and across a wide range of nonprofit human service organizations (Donaldson, 2004). A common criticism of the field is the lack of, and lack of sophistication of, evaluations of organizational development programs. Program theory-driven evaluation science, as described in this book, offers a feasible approach for evaluating and improving these types of organizational development interventions. Furthermore, many of the applications of program theory-driven evaluation science discussed in this book deal with programs concerned with the pursuit of improving worker productivity, health, and quality of work life.

Improving Organizational Effectiveness. Another potentially effective application of program theory-driven evaluation science is to improve the organizational effectiveness of organizations whose main purpose is to implement transformation technologies through programs. For example, many human service organizations, like the ones that are discussed later in this book, describe their mission, purpose, or why they exist through their transformation technologies (to prevent drug abuse, to promote healthy child-rearing practices, to mentor and educate tomorrow's leaders, to facilitate the reemployment of recently unemployed workers, to improve employment opportunities through technology training, etc.). To be recognized as effective, these organizations must be able to demonstrate that the clients they serve are indeed better off on the critical dimensions after receiving their services. It is well known that the success or failure of these programs, or transformation technologies, is highly dependent on the organization's ability to implement the services well, and that many organizations fail to implement programs well enough to achieve results (Donaldson, 2003). Program theory-driven evaluation science can be used to remedy this shortcoming by sorting out effective from ineffective services, as well as providing the information necessary for the organization to become and remain effective.

Applications

In chapter 4 through chapter 8, I illustrate how program theory-driven evaluation science was used to help develop and improve programs and a wide range of organizations participating in a large Work and Health Initiative. In this initiative, organizations were funded to develop new programs or to improve existing programs, and to develop or improve the organizations in which the programs were to be imbedded. As has been discussed in recent years (e.g., Sanders, 2001; Scriven, 2003), it is becoming more common for project funders to expect and require organizational leaders to use disciplined, professional evaluation (planned, systematic, funded evaluations adhering to professional guidelines and standards) to increase the likelihood of program and organizational success. That is, it is no longer the norm, nor acceptable, in many domains for organizational leaders to avoid or resist using professional evaluation to better their services, to prevent null or harmful effects from adversely impacting their clients, or to fail to rigorously account for the resources they receive from the supporters of their mission and organization. After all, don't most organizations owe this level of professionalism and accountability to their clients, employees, volunteers, and society at-large?

This book attempts to illustrate that program theory-driven evaluation science has the potential to help program and organizational leaders dramatically improve effectiveness. For example, program theory-driven evaluation science can help leaders:

1. Make better decisions about program and organizational direction.
2. Build knowledge, skills, and develop a capacity for evaluative thinking.
3. Facilitate continuous quality improvement and organizational learning.
4. Provide accountability to funders.
5. Justify the organization's value to investors, volunteers, staff, and prospective funders and supporters.

In addition, chapter 11 describes a proposal of how to use program theory-driven evaluation science to address a hypothetical evaluation problem presented and critiqued by scholars studying the differences between modern evaluation approaches (Alkin & Christie, 2005). In chapter 12, I explore a range of lessons learned from contemporary practice about using program theory-driven evaluation science to develop and improve people, programs, and organizations. Finally, a broader discussion of future directions for using evaluation science to promote human and social betterment is presented in chapter 13.

PURPOSE AND ORGANIZATION OF THIS BOOK

The purpose of this book is to provide a "state-of-the-art" treatment of the practice of program theory-driven evaluation science. It attempts to fill a serious void in the extant literature, namely a lack of detailed examples of program theory-driven evaluation science being implemented in "real world" settings (Donaldson, 2003; Weiss, 1997). That is, instead of relying on abstract theory or hypothetical examples to discuss this evaluation approach, an in-depth description of the nuances and results from a series of "authentic program theory-driven evaluations" from recent evaluation practice is presented. The presentation of each case will be organized around the concise three-step model for conducting program theory-driven evaluation science:

1. Developing program impact theory.
2. Formulating and prioritizing evaluation questions.
3. Answering evaluation questions.

First, chapter 2 and chapter 3 provide details about strategies for carrying out each of the three steps for conducting program theory-driven evaluation science. After discussing each of these steps, recent applications of program theory-driven evaluation science are presented in chapter 4 through chapter 9 to illustrate the process and potential products of this evaluation approach. Chapter 10 and chapter 11 discuss a second case and provide an example of an evaluation proposal based on a program theory-driven evaluation science framework. Challenges and lessons learned from the cases presented and the theory-driven evaluation literature more broadly are described in chapter 12. This discussion will include cross cutting lessons learned from the Work and Health Initiative, as well as from a range of other program domains. Finally, future directions for using evaluation science to improve society and the human condition are discussed in chapter 13. It is my hope that this book helps readers learn how to better use this powerful tool called program theory-driven evaluation science to prevent human suffering and to promote human welfare and achievement.

2

Developing Program Impact Theory

The purpose of this chapter is to describe the first step in the basic three-step process for conducting program theory-driven evaluation science:

1. Developing program impact theory.
2. Formulating and prioritizing evaluation questions.
3. Answering evaluation questions.

This chapter emphasizes the importance of conceptual framing in evaluation practice. A special emphasis is placed on clarifying the distinction between conceptual and empirical work in program theory-driven evaluation science. Next, a rather detailed discussion about the content of program impact theories is presented. Finally, the process of developing program impact theory is explored in some depth.

EFFICACY VERSUS EFFECTIVENESS EVALUATION

An important distinction to make in program evaluation practice today is whether the purpose of an outcome evaluation is to determine program efficacy or program effectiveness. In short, program efficacy evaluation typically determines whether a program works under ideal or researchlike conditions. Programs that do not turn out to be efficacious under highly controlled conditions should be abandoned or revised (if possible) before they are implemented more widely in society. On the other hand, evaluating programs being implemented for clients, service recipients, or consumers in "real world" school, health care, organizational, community settings and the like can be referred to as program effectiveness evaluation; for example, does the program make a difference in society? It could be argued that all programs should be subjected first to program efficacy evaluation, and if successful, subsequently implemented in the field and subjected to program effectiveness evaluation. However, this ideal is seldom realized, and programs very often bypass efficacy evaluation and are developed, implemented, and evaluated in the field. In fact, one might argue that the bulk of contemporary evaluation practice now involves program effectiveness evaluation.

I suspect that many of the recent debates and tensions, in the field of program evaluation at least, partially stem from the lack of clarity about the different challenges facing program efficacy versus program effectiveness evaluation (see Donaldson & Christie, 2005; Mark, 2003). For example, it is not uncommon to observe program effectiveness evaluations being designed with, or being judged by, standards and methods more appropriate for program efficacy evaluations. Although most modern evaluation approaches seem to be better suited for one or the other, efficacy or effectiveness, program theory-driven evaluation science strives to be equally effective across both domains. That is, the three-step process of (a) developing program impact theory, (b) formulating and prioritizing evaluation questions, and (c) answering evaluation questions is intended to be useful for both types of evaluation. However, it is important to underscore that the details (process, methods, data analysis, etc.) of each step may differ rather dramatically across efficacy versus effectiveness evaluations.

Many of the writings on theory-driven evaluation to date have been oriented toward issues most relevant to efficacy evaluation (cf. Chen, 1990; Donaldson, 2003; Rossi et al., 2004; Weiss, 1997). However, most practicing evaluators today need information about how best to conduct effectiveness evaluations. Effectiveness evaluations must deal with a wide range of issues that stem from dynamic, participatory interactions with stakeholders, and from having less control over characteristics of the participants, service providers, and program and organizational contexts. The strategies and applications presented in this book represent an attempt to fill this gap in the literature. You will read about strategies for working collaboratively with stakeholders and observe from authentic cases how program theory-driven evaluation science has been implemented in real world effectiveness evaluations.

CONCEPTUAL FRAMING

One of the key lessons from the past three decades of evaluation practice is that effectiveness oriented program evaluations rarely satisfy all stakeholders' desires and aspirations. Unrealistic or poorly managed stakeholder expectations about the nature, benefits, and risks of evaluation quickly lead to undesirable conflicts and disputes, lack of evaluation utilization, and great dissatisfaction with evaluation teams and evaluations (see Donaldson, 2001b; Donaldson, Gooler, & Scriven, 2002). A major contribution of participatory program theory-driven evaluation science is its emphasis on working with relevant stakeholders from the outset to develop a common understanding of the program and to develop realistic expectations by tailoring the evaluation to meet agreed-on values and goals. That is, a conceptual framework is developed and used to tailor the usually more costly empirical portion of the evaluation. The empirical work is focused on answering the most important questions. Evaluation questions of most interest to relevant stakeholders are answered using the most rigorous methods possible given the practical constraints of the evaluation context.

The first task of a systematic program theory-driven evaluation is to develop a conceptual framework or program theory of how a program intends to solve the problem of interest (i.e., meet the needs of its target population). Three common definitions of program theory capture the essence of how I use this term throughout this book:

- The construction of a plausible and sensible model of how a program is supposed to work (Bickman, 1987).
- A set of propositions regarding what goes on in the black box during the transformation of input to output; that is, how a bad situation is transformed into a better one through treatment inputs (Lipsey, 1993).
- The process through which program components are presumed to affect outcomes and the conditions under which these processes are believed to operate (Donaldson, 2001a).

It is highly desirable if program theory is rooted in, or at least consistent with, behavioral or social science theory or prior research (see Donaldson, Street, Sussman, & Tobler, 2001). However, often sound theory and/or research is not available for the problem of concern. If this is indeed the case, other sources of information are used to develop program theory, including implicit theories held by those closest to the operation of the program, observations of the program in action, documentation of program operations, and exploratory research to test critical assumptions. The specific content and process for developing program theories to improve the quality of program evaluations is now described in some detail.

THE CONTENT OF PROGRAM THEORY

There are now a wide variety of ways to represent a program theory. However, there does not seem to be a consensus in the literature today about how best to diagram or describe it. For example, program theory is sometimes described by a logic model, program model, theory of change, cause map, action theory, intervening mechanism theory, and the like (see Donaldson, 2003). In general, however, most program theory-driven evaluations provide some type of figure, diagram, or pictorial version of a program theory that is accompanied by a more detailed written description. Program theories (and programs) may vary considerably in terms of complexity. Although evaluators and evaluation teams should typically develop and consider multiple program theories, some representing the possible complexity of the situation, I recommend striving for a parsimonious version to use with stakeholders in effectiveness evaluations. A parsimonious program theory is focused on the "main active ingredients" that are presumed to lead to the desired outcomes, and the key conditions under which they are believed to operate.

Rossi et al. (2004) introduced a scheme for representing program theory that helps sort out differences between program process and program impact theory. Figure 2.1 presents this overview diagram for representing program theory.

The *program process theory* maps out the inner workings of the implementation of the program. That is, it attempts to diagram and articulate the programs' assumptions about how it is reaching its target population, and describes the nature of the interactions between the

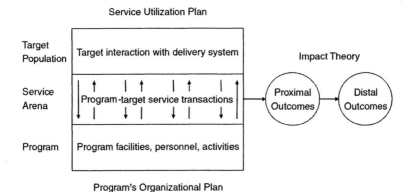

Figure 2.1. Overview of program theory. From P. H. Rossi, M. W. Lipsey, and H. E. Freeman, *Evaluation* (7th ed.), p. 80, copyright © 2004 by Sage Publications. Reprinted by permission of Sage Publication, Inc.

Figure 2.2.   Program impact theory example.

target population and the program in its *service utilization plan*. The program's *organizational plan* maps out the facilities, personnel, activities, administration, and general organization needed to support providing these services to the target population. Program process theory can be integral to program planning and can be used to help structure evaluation questions about the actual program implementation in real-world effectiveness evaluation settings.

Program impact theory, on the other hand, describes the cause-and-effect sequences that link the program services and activities to proximal and distal outcomes. A typical program impact theory illustrates how a program is expected to result in important desired outcomes, solve the problems of interest, or meet the needs of its target population. For example, Figure 2.2 illustrates a simple, basic two-step program impact theory.

This illustrates that workers who participate in a workplace health promotion program are expected to obtain higher levels of fitness (proximal outcome), which in turn is expected to improve their job performance (distal outcome). The problem of interest in this example is low job performance. The program is expected to solve the problem by improving employee fitness. Put another way, this program rests on the assumption that employees need to raise their fitness level in order to perform well at work. This identified need could be based on a hunch of the designers of the program, or it could have been the primary finding identified in a systematic needs assessment. The main point to emphasize here is that program impact theory illustrates the expected results of a program, in contrast to program process theory, which describes the nature of the program itself. Before I move on to elaborate on these central concepts, I briefly compare a typical logic model with the representation of program theory.

A very popular tool for depicting program logic and theory in recent years is called a *logic model*. A logic model is an adaptable tool that is now being used across a range of evaluation approaches to assist with program planning, design, and evaluation (see Funnell, 1997). There are many approaches to logic modeling (e.g., McLauglin & Jordan, 1999;

Figure 2.3.   How the standard logic model framework relates to program theory.

Renger & Titcomb, 2002). However, one of the most standard or universal formats today involves mapping out a program's inputs, activities, outputs, initial outcomes, intermediate outcomes, and longer term outcomes. Figure 2.3 illustrates how the standard logical model framework relates to the program theory scheme presented in Figure 2.1.

The inputs, activities, and outputs described in a logical model would be part of the program process theory in the program theory representation. The initial, intermediate, and long-term outcomes would be part of the program impact theory.

Both program process theory and program impact theory can be crucial for helping to design and conduct sound program evaluations. Program process theory is especially useful for understanding the design of a program, and for framing evaluation questions that might help determine if a program is actually implemented as intended. If a program is not implemented as intended, outcome evaluations can be misleading. For example, it is possible to attribute disappointing results of a program evaluation to the nature of the program design (the program is not a good idea after all), when they should be attributed to the fact that the program was not implemented properly (e.g., participants did not attend, services were poor, etc.). In general, it is important to make sure a program has been implemented properly, as specified in the program process theory, in advance of evaluating outcomes described in the program impact theory. This is similar to using a manipulation check in basic experimental research to make sure the experimental manipulation worked before interpreting the results. It could be said that program impact theory is the main event (findings in basic research), and is based on the assumption that a program has been effectively delivered to the target population. The program process theory helps us frame questions and gather data to confirm that this critical assumption about program implementation (the manipulation check) is correct.

Program impact theory is typically much more detailed and complex in practice than the overview portrayed in Figure 2.1. The main components of most program impact theories are the program, proximal or immediate outcomes, distal or more long-term outcomes, and sometimes conditioning or moderating factors that influence the strength and/or direction of the paths between the other factors (see Donaldson, 2001a). Some stakeholders prefer pictures that use circles connected by arrows, others prefer ellipses, rectangles, squares, and so forth. Although there are not universally accepted conventions here, it is important that the picture is believed to be a meaningful depiction of how the program is presumed to operate by the stakeholders involved in an effectiveness evaluation. Some critics believe the popular use of logic modeling, often used as a tool to develop program theory (Funnell, 1997), is too simplistic and restricts program theory to linear relationships. However, this is a myth because there are no such restrictions; program theories may reflect a wide range of possible relationships (including nonlinear, interactive, and recursive relationships among others) between constructs (see Donaldson, 2003; Lipsey, 1990; Lipsey & Pollard, 1989).

The decisions of how best to represent the program of interest are made in collaboration with the stakeholders in a participatory program theory-driven evaluation. Although there are endless possibilities of how to pictorially represent program impact theories, the most widely recognized and used approach might be considered variable oriented. A *variable-oriented program impact theory* consists of relationships among variables or constructs typically described as one of three basic types: (a) direct or main effects, (b) indirect or mediator relationships, or (c) moderator relationships (Donaldson, 2001a; Lipsey & Pollard, 1989). This approach is used frequently in both efficacy and effectiveness evaluation and research, and is widely supported or consistent with common methods and analytic techniques. The variable-oriented approach to representing program impact theory provides an important bridge between research and evaluation, efficacy and effectiveness studies, and theory and evaluation practice. I now describe how each type of relationship is used to represent the relationships between constructs in a program impact theory.

Direct Effect. In the past, the standard and most common conceptualization of program theory was known as a direct effect. That is, a program was typically conceived to affect an outcome or outcomes (see Fig. 2.4). The focus of most program developers was to include as many activities as possible in the program in an effort to make sure outcomes were in fact improved. The task of the evaluator was to demonstrate whether or not a program did in fact improve desired outcomes. Reviews of the program evaluation literature have often concluded, in most studies, that programs are conceptualized as undifferentiated "black box"

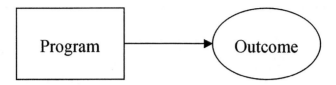

Figure 2.4.   Direct effect program conceptualization.

treatment packages, and only report information about direct effects (Lipsey, 1988; Lipsey, Crosse, Dunkel, Pollard, & Stobart, 1985; Lipsey & Wilson, 1993).

One main limitation of the direct effect conceptualization and evaluation is that little is learned when there is no overall program effect. For example, the researcher is not able to disentangle the success or failure of program implementation from the validity of the conceptual model on which the program is based (Chen, 1990; Donaldson, 2003, 2005). Similarly, there is no way to sort out which components of a program are effective, and which are ineffective or actually harmful. Another serious problem with the direct effect conceptualization is that behavioral interventions only have indirect effects. Hansen and McNeal (1996) call this the *Law of Indirect Effect*:

> This law dictates that direct effects of a program on behavior are not possible. The expression or suppression of a behavior is controlled by neural and situational processes, over which the interventionist has no direct control. To achieve their effects, programs must alter processes that have the potential to indirectly influence the behavior of interest. Simply stated, programs do not attempt to change behavior directly. Instead they attempt to change the way people think about the behavior, the way they perceive the social environment that influences the behavior, the skills they bring to bear on situations that augment risk for the occurrence of the behavior, or the structure of the environment in which the behavior will eventually emerge or be suppressed. The essence of health education is changing predisposing and enabling factors that lead to behavior, not the behavior itself. (p. 503)

Therefore, the direct effect conceptualization alone is limited and is usually not enough to represent program impact.

Mediation Models.   In contrast, using an indirect effect conceptualization of a program can dramatically improve understanding about a program's functioning. I illustrate this point first by only modestly adding one additional variable to the direct effect conceptualization shown in Figure 2.4. Figure 2.5 illustrates a basic two-step mediation model. This

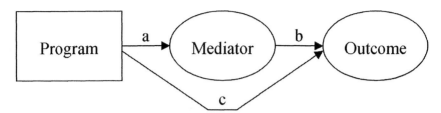

Figure 2.5.   Indirect effect program conceptualization.

model shows that the program affects the mediator (Path A), which in turn affects the behavioral outcome (Path B). Path C represents the residual direct effect. Theoretically, if Path C is zero, the mediator variable is the lone cause of the outcome. If Path C is nonzero, there are believed to be additional mediators that explain the link between the program and the outcome. Please note that the label *mediator* is synonymous with proximal outcome in the program theory scheme in Figure 2.1, and an initial outcome in the logic model depiction in Figure 2.3. It is also important to note that with just this additional variable, an evaluator is now in a position to determine (a) whether or not the program was effective enough to alter its target, and (b) whether or not the program is aimed at the right target (the validity of the conceptual model; that is, does the mediator variable in fact improve the outcome variable?).

Most programs are considered to be multiple-component interventions, which are more accurately conceptualized to contain multiple mediator variables (see Fig. 2.6). In practice, it is probably most often the case that program components reflect the program developer's view of how to optimize the chances of obtaining desired outcomes, rather than a combination of components shown to be effective (cf. Hansen, 1993). West and Aiken (1997) described fundamental tensions between program developers and evaluators associated with the development of multicomponent programs. Put simply, although program developers are primarily concerned with maximizing the efficacy of the entire program, evaluators have become increasingly concerned with how each component affects targeted mediators or risk factors and produce the effects on the desired outcomes.

I believe more complex multiple mediation model conceptualizations of programs are usually necessary to accurately reflect the multicomponent nature of most modern programs. These conceptualizations can enhance the program development and evaluation process by more precisely identifying the program characteristics that are presumed to influence each target mediator. Pilot empirical work can then be

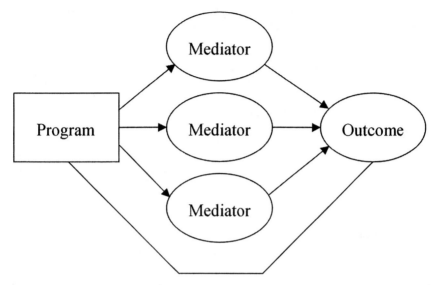

Figure 2.6.   Multiple-mediator model.

conducted to estimate whether it is likely that each individual program component will have a large enough effect on the targeted mediators to have an effect on desired outcomes when the program is actually implemented. Carefully selecting mediators to explain and maximize the effects of the program should be the common goal of program developers and evaluators.

Another important benefit of conceptualizing a program in this way is that it exposes, often implicit, theoretical program mechanisms; that is, the paths going from the mediators to the outcomes represent the conceptual foundation of the program. In some cases, the link between the assumed risk factor (mediator) and the outcome has been well established in prior research. In other cases, there is considerable ambiguity about the links. It is not uncommon for program developers who use this approach to dramatically modify the program because the original conceptual links appear very unrealistic once they are exposed. It is important to keep in mind that the magnitude of change in a behavioral outcome a program can produce is directly limited by the strength of relationships that exist between mediators and outcomes. Hansen and McNeal (1996) describe this as *The Law of Maximum Expected Potential Effect*:

> The magnitude of change in a behavioral outcome that a program can produce is directly limited by the strength of relationships that exist between

mediators and targeted behaviors. The existence of this law is based on the mathematical formulae used in estimating the strength of mediating variable relationships, not from empirical observation, although we believe that empirical observations will generally corroborate its existence. An understanding of this law should allow intervention researchers a mathematical grounding in the selection of mediating processes for intervention. An added benefit may ultimately be the ability to predict with some accuracy the a priori maximum potential of programs to have an effect on targeted behavioral outcomes, although this may be beyond the current state-of-the-science to achieve. (p. 502)

Therefore, it is critical to make sure that the program is aiming at the right targets (mediators) if it is going to have any chance of achieving outcomes.

Path Moderators.   In general, a *moderator* is a qualitative (e.g., gender, ethnicity) or quantitative (e.g., amount of outside practice, degree of motivation) variable that affects the direction and/or strength of the relationships between the program and mediator, or mediator and the outcome (see Baron & Kenny, 1986). Figure 2.7 illustrates that the moderator conditions or influences the path between the program and the mediator. This means that the strength and/or the direction of the relationship between the program and the mediator are significantly affected by the moderator variable. This type of moderator relationship is of primary importance in program development. Program developers can benefit greatly from considering if potential moderator variables such as participant characteristics, provider characteristics, characteristics of the setting of program

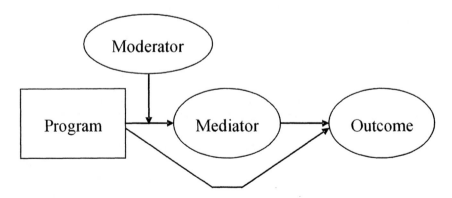

Figure 2.7.   Moderator of program–mediator relationship.

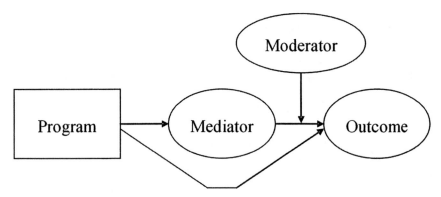

Figure 2.8. Moderator of mediator–outcome relationship.

implementation, strength of the programs, and program fidelity significantly influence the program's ability to affect target mediators.

Figure 2.8 illustrates that the moderator conditions the path between the mediator and outcome. This type of relationship addresses the generalizability of the presumed program mechanism. For example, is the relationship between the mediator and the outcome the same for (a) females versus males, (b) inner city children versus children living in the suburbs, or (c) Latinos, African Americans, Asian Americans, and European Americans? Although it is most often the case that the relationships between the mediators and the program are presumed to generalize across all participants, in some cases, it is possible that participant characteristics condition the relationships between the mediators and the outcomes. It is important to identify significant moderators of all paths when conceptualizing how a program is presumed to work. These relationships may be critical to understanding the program effects, or the lack thereof, later when the program is being evaluated.

In summary, the variable-oriented approach has been the most commonly used approach in program theory-driven evaluation science to represent program theories. The foundation of the variable approach consists of three basic relationships (direct, mediator, and moderator relationships). A variety of more complex program impact theories are possible using the basic building-block relationships as a foundation (i.e., direct, indirect, and moderator). For example, multiple mediation links may be posited for linking a program with its outcomes. Once each relationship is specified, a thoughtful and probing analysis of each arrow depicting the relationships is required to examine program impact theory in more detail. I discuss ways to conduct that further examination of the arrows between constructs as part of the next section on the process of developing program impact theory.

THE PROCESS OF DEVELOPING PROGRAM IMPACT THEORY

Characteristics of the evaluation and the program affect the nature of the process for developing program impact theory. For example, two characteristics that heavily influence how an evaluator or evaluation team develops program impact theory in practice include whether the evaluation is focused on program efficacy or effectiveness, and the maturity of the program (e.g., whether the program is established or new). For example, efficacy evaluations of new programs typically rely primarily on prior theory and research in the program domain to develop program impact theory (see Donaldson et al., 2001). A common scenario here is that a theorist(s) or researcher(s) develops a program based on what they have learned previously about the problem or phenomena under investigation. Given their theoretical or knowledge base from prior research, the question is, can they design an intervention or program that will prevent or solve the problem(s) of interest under somewhat controlled or researchlike conditions (i.e., can the program have an effect under "ideal conditions"?)? The program impact theory in this case specifies or hypothesizes how they plan to remedy the undesirable situation.

In contrast, effectiveness evaluation of an existing program is concerned with: Does a program work under "real conditions"? For example, does the program solve the problem(s) of interest when it is being delivered in a human service organization, school, community setting, or for-profit corporation (see Fitzpatrick, 2002). Under these conditions, the evaluator or evaluation team is typically required to work with stakeholders to make implicit program theory explicit and testable. Fortunately, it is often possible and highly desirable in this type of situation to base program theory on multiple sources of information such as (a) prior theory and research in the program domain; (b) implicit theories held by those closest to the operation of the program (program personnel such as health educators or other human service providers); (c) observations of the program in action; (d) document analysis; and in some cases, (e) exploratory research to test critical assumptions about the nature of the program. This process seems to work well when evaluators and stakeholders approach this as a highly interactive and nonlinear exercise (cf. Donaldson & Gooler, 2003; Fitzpatrick, 2002).

The first program theory-driven evaluation presented in this book (chap. 4) involves an effectiveness evaluation of an existing program that has been exported to, and implemented in, a new geographic region. The program conditions and participants were very different than in previous implementations and evaluations of the program. Although the initial versions of the program impact theory were highly influenced by prior theory, research, and evaluations, implicit theories, observation, and exploratory research were used to refine program impact theory development.

The other three program evaluations (chap. 5, chap. 6, and chap. 7) are effectiveness evaluations of new programs. Prior theory and research was rather sparse in these cases, so program impact theory development relied much more on making implicit theory explicit, documentation, observation, and to a lesser extent exploratory research. Finally, in chapter 8, a rather detailed account of how I proposed to conduct an effectiveness evaluation for an ongoing or rather mature program. In each of these applications, some general strategies and principles were followed to help stakeholders fully conceptualize their programs prior to specifying evaluation questions, design, or methods. I now briefly outline some of the general strategies that were used to develop program impact theory in the effectiveness evaluations described in this book.

Engage Relevant Stakeholders. In each case, our first objective was to engage as many relevant stakeholders as possible in discussions about the nature of the program of interest. In theory, we aspired to be as inclusive as possible. However, in practice, we always encountered obstacles such as lack of interest, availability, time, and/or resources that prevented us from including all relevant stakeholders. The strategy we used to deal with these obstacles was to try to get as many representatives as possible from the various constituencies (program leadership, staff, the funder, clients, etc.) to participate in meetings with the evaluation team to discuss the nature of the program.

The evaluation team tried to convey genuine interest in understanding the program and the stakeholders' values, beliefs, and experiences. It was made clear that we wanted to gain a deep appreciation of the work going on before even thinking about, let alone discussing potential measurement or an evaluation design. Most stakeholders seemed to be relieved that we did not come across as presumptuous, and were planning to take the time to get to know them and their program before suggesting how to measure and evaluate.

Develop a First Draft. Much of our initial discussions focused on what they were hoping the program would accomplish. Although they were describing this in their own words, we were trying to identify outcomes of interest in our private notes. Once we felt we had some initial indication of the expected long-term outcomes, we asked the stakeholders to discuss the process through which they believed the outcomes would be produced. This type of discussion typically led us to identify some possible short-term outcomes and potential moderator relationships. It is important to note that this conversation was in common everyday language, and we did not ask them to articulate constructs, relationships between constructs, or to describe their program impact theory. Our objective was to create a first draft of a program impact theory based on their rich description of the program.

Present First Draft to Stakeholders. The evaluation team worked together independently of the stakeholders to produce a first draft of a program impact theory based on the initial discussions. This process typically involved identifying key components of the theory—mediators, moderators, and outcomes—and then discussing the links between them. We typically worked right to left first; that is, based on our conversations, we determined a set of key desired outcomes. Next, we asked ourselves, how do they expect the program to produce these desired outcomes, which led us to identify mediators (or short-term outcomes) and potential moderators. We then worked through the model again as a group (left to right, right to left) to make sure the links made sense based on our conversations.

Meetings were arranged for the evaluation team to discuss the first draft of the program impact theory with the stakeholder representatives. Sometimes this was done in face-to-face meetings; other times this was done by a telephone conference call. All participants were mailed or e-mailed a copy of the program impact theory in advance so they would have time to think about the accuracy of the initial draft. Most often these meetings led to some revisions of the initial program impact theory.

Plausibility Check. Once the evaluation team and stakeholders reached general agreement about the program impact theory from the stakeholders' point of view, the evaluation team began to examine the plausibility of the links in the model. This was most often done by reviewing prior research and evaluations relevant to the program impact theory, with a particular focus on assessing the plausibility of each link. It was very common to find implausible links based on what was currently known about the constructs of interest. This led to more discussions with the stakeholders to determine whether additional revisions to the program impact theory were needed. Sometimes this exercise suggested that serious program improvements were needed to increase the chances that the program would produce desired outcomes. Other times we discovered "magical or overly optimistic thinking," which led the stakeholders to get more realistic about what they expected from the program in term of outcomes. At the end of the discussion about plausibility, the evaluation team attempted to gain a consensus about a working program impact theory.

Probe Arrows for Model Specificity. The evaluation team, working independently from the stakeholder groups, again examined the nature of the program impact theory at a deeper level of detail for each of the effectiveness evaluations. The goal of this step was to finalize the model so that it would be ready to use for generating specific evaluation questions. In general, parsimonious models focused on the "main active ingredients" needed to produce key desired outcomes tended to be the most useful for designing sensitive and valid program evaluations (see Donaldson, 2003; Donaldson & Gooler, 2002). For example, the timing and nature of each arrow were examined carefully. Common questions asked included: Are the relationships represented by the arrows expected to be linear? What time lag do the arrows represent (does the program affect the outcome immediately, 1 week later, 6 months later, etc.)?

Lipsey (1990) provided some useful figures for thinking about the relationships between a program and its expected outcomes. Figure 2.9 illustrates four possible program-effect decay functions. The first pattern (A) shows the possibility that the program has a large impact on the outcome of interest, and that there is no decay of that effect over time. This is every program developers dream result but probably a rare relationship in most domains of effectiveness evaluation. The second pattern (B) illustrates the possibility that a program affects an outcome some time after the program has been delivered. A delayed effect is challenging to capture unless evaluators have the will and budget to measure expected outcomes over an extended period of time. It is common for null results in effectiveness evaluation to be attributed to the belief that the expected positive benefits likely occurred after the evaluation was completed (sometimes possible, other times an excuse for lack of success). The next pattern (C) illustrates that a program had an immediate effect on an outcome, but the effect rapidly decayed. Skeptics of programs and trainings that use an immediate posttest (same-day smiley sheets) suspect the effects of many programs are very short-lived like this pattern would suggest. Finally, pattern D illustrates an early effect with slow decay over time. This is a realistic pattern for successful programs, and suggests the value of a booster program to help retain the integrity of the initial intervention. These are just a few examples of what an arrow between constructs in a program impact theory might actually represent in practice. The evaluation team needs to think carefully about these and other possible patterns

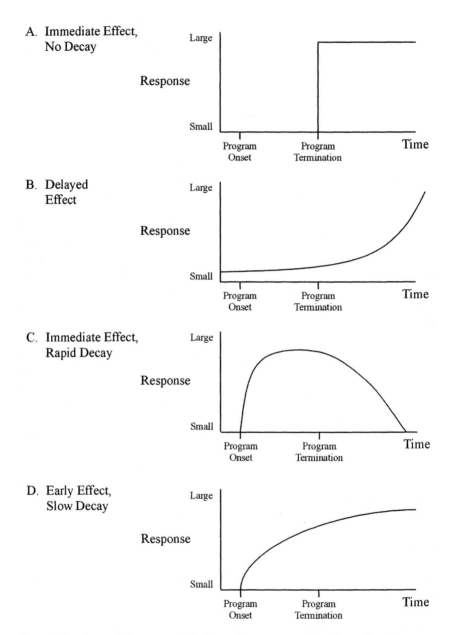

Figure 2.9. Some different possible forms for program-effect decay functions. From M. W. Lipsey, *Design Sensitivity*, p. 149, copyright © 1990 by Sage Publications. Reprinted by permission of Sage Publications, Inc.

so that the timing of the measurement, and the evaluation design in general, is sensitive enough to reflect the nature of the phenomenon under investigation. Stated another way, it is easy to measure outcomes at the wrong time in an effectiveness evaluation, which can lead to misleading or incomplete evaluation conclusions.

Let's take one more example of how the nature of the arrows in a program impact theory might be examined in more detail. Some programs come in different doses; sometimes this is because of differential program implementation across sites, individuals attend or participate at different rates, and other times this occurs by design. Lipsey (1990) provided five examples of different dose-response functions that could represent the arrow or relationship between a program and expected outcome (see Fig. 2.10). Pattern A suggests a moderate dose produced a large effect. Pattern B shows the possibility that only a strong dose of the program will produce the outcomes. Pattern C illustrates the possibility that the program is so relevant that it affects the outcome even with weak doses. If this pattern were true, it would potentially suggest this was a cost-effective program that only needed to be delivered in a small dose to produce the desired benefit. Pattern D suggests the possibility for large effects for small and large doses, but no effect for moderate doses. Pattern E suggests that the optimal dose is moderate, where too much or too little of the program fails to result in the outcome of interest. These examples are intended to illustrate that the arrows in a program impact theory may not be simple linear relationships and could represent one of a wide variety of possible patterns.

Finalize Program Impact Theory (Step 1).   The evaluation team explored each arrow in detail in the cases presented in this book. Often, minor modifications to the program impact theories were made at this step to provide more specificity and precision. Suggested adjustments were presented to stakeholders for discussion and final approval. As you will see, some stakeholder groups preferred to use a parsimonious program impact theory as the guiding framework, and to build the specifics explored at the previous step into the evaluation questions and design phase of the process. In general, we recommended this approach so that the common representation of the program impact theory did not become overly complex and overwhelming for the diverse group of stakeholders. Other stakeholder groups preferred to include more detail and complexity in their guiding program impact theory (e.g., the Health Insurance Policy Program, chap. 8). In summary, the evaluation team facilitated this step but the stakeholders ultimately decided on the final

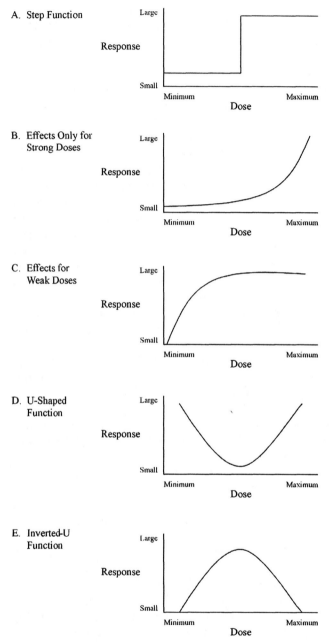

Figure 2.10. Some different possible forms for dose-response functions. From M. W. Lipsey, *Design Sensitivity*, p. 156, copyright © 1990 by Sage Publications. Reprinted by permission of Sage Publications, Inc.

representation of the program impact theory that would be used to generate, prioritize, and answer evaluation questions.

A major obstacle that we typically encountered during the development of program impact theories was the natural desire of the stakeholders and evaluation team members to worry about evaluation questions, methods, measures, costs and the like while they were trying to conceptualize the program. We have found this can seriously constrain and hinder the development of useful program impact theories. I believe it is important to emphasize that the development of program impact theories should be done independently of, and isolated from, the empirical evaluation design. This allows stakeholders and evaluators to fully explore the nature of the program free of constraints. Once the program is fully conceptualized and a working program impact theory is adopted, the evaluation team and stakeholders begin the question formulation phase (Step 2). That is, the program impact theory is used to guide the stakeholders as they formulate and prioritize evaluation questions as the second step of the program theory-driven evaluation science process.

CONCLUSION

An overview of how to develop program impact theory was provided in this chapter. A specific emphasis was placed on showing how program process and program impact theory can help evaluators and stakeholders develop a common understanding of a program. The process of developing this common understanding can help build communication and trust between the stakeholders and the evaluation team. It also typically leads to more informed evaluation questions and more sensitive and valid evaluation designs. Stated another way, this first step of developing program impact theory forces evaluators to think more deeply about a program before moving to the more familiar and common tasks of designing measures, data collection strategies, planning for data analysis, and the like. The program impact theory developed in Step 1 of the program-theory driven evaluation science process will be used to guide Steps 2 and 3, which are discussed more fully in the next chapter.

3

Formulating, Prioritizing, and Answering
Evaluation Questions

The purpose of this chapter is to briefly describe the second and third step in the basic three-step process for conducting program theory-driven evaluation science:

1. Developing program impact theory.
2. Formulating and prioritizing evaluation questions.
3. Answering evaluation questions.

This chapter emphasizes the importance of using the program impact theory developed in Step 1 to facilitate discussions with stakeholders about formulating, prioritizing, and answering potential evaluation questions. First, I discuss ways that an evaluation team can help stakeholders formulate a wide range of evaluation questions. Next, a process for prioritizing the evaluation questions is explored. Finally, issues related to designing evaluations that provide credible answers to key evaluation questions are addressed.

FORMULATING EVALUATION QUESTIONS

I strongly encouraged evaluators in chapter 2 to develop program impact theory free of the constraints of specifying evaluation questions, or which methods would be desirable and/or feasible for testing program impact theory. Only after program impact theory has been fully developed do I recommend the evaluator or evaluation team, in collaboration with the relevant stakeholders, begin listing potential evaluation questions.

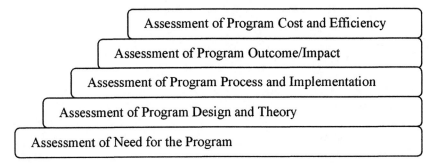

Figure 3.1. The evaluation questions hierarchy. From P. H. Rossi, M. W. Lipsey, and H. E. Freeman, *Evaluation* (7th ed.), p. 140, copyright © 2004 by Sage Publications. Reprinted by permission of Sage Publications, Inc.

The goal of the evaluation team should be to help the relevant stakeholders generate a solid list of possible evaluation questions that could be explored using empirical evaluation procedures.

Most evaluation textbooks provide an extensive discussion of a range of questions that can be asked in a typical program evaluation (cf., Chen, 2005; Fitzpatrick, Sanders, & Worthen, 2004; Mark et al., 2000; Rossi et al., 2004; Weiss, 1998). For example, Rossi et al. (2004) provided the evaluation question hierarchy displayed in Figure 3.1. This organizing framework illustrates the types of questions one might ask about:

- The need for program services.
- The program's conceptualization or design.
- Program operations and service delivery.
- Program outcomes.
- Program cost and efficiency.

Similarly, Weiss (1998) offered five categories of questions typically asked about programs that include:

- Program process.
- Program outcomes.
- Attributing outcomes to the program.
- Links between process and outcomes.
- Explanations.

Scriven (1991) provided the Key Evaluation Checklist (KEC) that outlines a range of issues and questions than can be explored in a program evaluation. For example, the KEC guides evaluators to explore questions about the:

- Nature of the evaluand (what is be evaluated).
- The background and context.
- Consumers.
- Resources.
- Values.
- Proccesses.
- Outcomes.
- Costs.
- Comparisons.
- Generalizability.
- Significance.
- Recommendations.
- Reporting.
- Metaevaluation.

Furthermore, Scriven (2003) added "The Something More" list to illustrate that evaluators need to draw on more than typical social science methods to formulate and answer key evaluation questions and to conduct professional evaluation.

As you might imagine, there are hundreds of possible questions that could be asked in any given program evaluation (Weiss, 1998). The role of the evaluation team at this step is to educate the stakeholders about the types of questions that could be asked about their program. The program impact theory is a useful guide for helping stakeholders formulate specific questions about the nature and context of their program. In chapter 11, sample questions are provided to illustrate how to formulate potential questions about an educational program. Some of these questions include:

- Are the intended curricula being delivered with high fidelity to the intended students?
- Are there students or families with needs that the program is not reaching?
- Are the desired short-term outcomes (mediators) being achieved?
- Are the desired longer term or ultimate outcomes of concern being achieved?
- Does the program have any adverse side effects?
- Are some recipients affected more by the program than others (moderator effects)?
- Does the program work better under some conditions than others (moderator effects)?
- Are the resources being used efficiently?
- Is the cost reasonable in relation to the benefits?
- Would alternative educational approaches yield equivalent or more benefits at less cost?

Many other examples of specific evaluation questions developed in the context of program impact theory appear throughout the subsequent chapters. As you see in the evaluations described in this book, the formulation of possible evaluation questions can be a somewhat lengthy process, and involve multiple meetings, teleconferences, and e-mail exchanges with stakeholders. However, it is highly important to invest time and effort in the process at this crucial step. Keep in mind the quality of any professional program evaluation rests heavily on formulating important and useful evaluation questions.

PRIORITIZING EVALUATION QUESTIONS

Once program impact theory has been used to identify a wide range of potential evaluation questions, the evaluation team and stakeholders attempt to prioritize these questions so that it is clear which questions are of most value. Differences of opinion about the value of each question across the stakeholder groups are noted, and factored into final decisions about which questions to answer.

In an ideal world, the entire program impact theory would be tested in the most rigorous fashion possible, and all valued evaluation questions would be answered. However, in most evaluations, only a subset of questions and components of a program theory can be evaluated due to time, resource, and practical constraints. Identifying and prioritizing the most important evaluation questions can prevent "paralysis in evaluation" (i.e., deciding to wait, or not to evaluate at all, because you cannot evaluate the entire program or set of evaluation questions). Recognizing that some key questions can be addressed at this time, even though some questions, or many others, will not be examined, facilitates the evaluation process to move forward.

A helpful approach for prioritizing a list of possible evaluation questions is to first ask stakeholders: How would a sound answer to the question of interest help the program? This forces stakeholders to think about each evaluation question in terms of its value to the overall effort to prevent or solve the problems of interest. The answer to some questions may be obvious and consequently add little value. Other questions may lead to answers that would shed light on how to solve a problem that is lowering the performance or quality of life of the target populations. Extensive discussions along these lines with the relevant stakeholders typically result in a clear, short list of questions that are likely to add the most value to the project.

Once a short list of valued evaluation questions is developed, the evaluation team should facilitate a process to determine how much it will cost to answer each question. At this point, evaluation design feasibility and stakeholder views about what type of evidence would provide credible answers become a major concern. In Step 3 of the program theory-driven evaluation science process, these issues are tackled head on. The list of valued evaluation questions generated in Step 2 is further refined in Step 3, so that the final set of evaluation questions pursued in the evaluation is reasonable, appropriate, valuable, and answerable (see Rossi et al., 2004).

ANSWERING EVALUATION QUESTIONS

In many respects, program theory-driven evaluation science is method neutral, and creates a superordinate goal that helps evaluators get past old debates about which methods are superior in program evaluation (e.g., the quantitative/qualitative debate; Reichhardt & Rallis, 1994). That is, from the contingency point of view, the theory-driven approach argues that quantitative, qualitative, or mixed methods designs are neither superior nor applicable in every evaluation situation (Chen, 1997). Instead, methodological choices are informed by program impact theory, specific evaluation questions ranked in order of priority, and practical constraints (Donaldson & Gooler, 2001; Donaldson & Lipsey, 2006). Therefore, the final step in program theory-driven evaluation involves determining what type of evidence is needed to answer questions of interest with an acceptable level of confidence.

The details of this step vary considerably across program evaluations. In some cases, the stakeholders will accept nothing short of evidence based on a large-scale randomized controlled trial (RCT). Whereas, in other cases, rich descriptions developed through qualitative methods are preferred over experimental designs. Further, unless data collection resources are plentiful, compromises are made to determine the most convincing design within resource and practical constraints. Unfortunately, there are no simple guidelines for making these decisions, but many useful resources now exist that describe the wide range of methods and designs that can be used to answer evaluation questions in contemporary practice (e.g., Chen, 2005; Davidson, 2005; Donaldson & Scriven, 2003b; Fitzpatrick et al., 2004; Levin & McEwan, 2001; Mark et al., 2000; Patton, 2001; Rossi et al., 2004; Scriven, 1991; Weiss, 1998). Suffice it to say here, many factors interact to determine how to collect the evidence needed to answer key evaluation questions. Some of these factors include design feasibility issues, the resources available, stakeholder preferences, and evaluation team expertise and preferences. The role of the evaluation team at this stage is to present options to the stakeholders about evaluation designs and methods that could be used to answer their most highly valued questions.

Next, stakeholders review the options presented by the evaluation team for collecting data and building evidence to answer the evaluation questions they value most. Each option has time, cost, and feasibility considerations in relation to the context of the program evaluation. It is not uncommon to encounter bitter disputes among stakeholders about which design or method is likely to produce the best evidence, taking into account cost and feasibility. This is the point at which hard decisions must be made about which questions to answer, and about what type of evidence to collect to answer those questions. In short, the evaluation team does the best job possible of presenting the strengths and weaknesses of each option in light of guiding principles (American Evaluation Association Guiding Principles, 2004) and evaluation standards (Joint Committee Standards for Educational Evaluation, 1994), and it is then up to stakeholders to decide which questions to pursue and which methods to use to answer those questions. The decisions made at this point typically narrow the list of questions that will be answered, and further refine the list of priorities set at Step 2.

Finally, it is advisable that the evaluation team and stakeholders agree to a specific contract for conducting the empirical portion of the evaluation. Proposing and discussing a detailed evaluation contract with the stakeholders is a great tool for setting realistic expectations about what an evaluation can actually deliver to stakeholders, and for determining roles for ensuring that evaluation findings and lessons learned will be disseminated and used to improve the lives of the participants, programs, and organizations of interest (Donaldson, 2001b). Sometimes data collection decisions made to address one evaluation question, with only minor additions, can be used to answer other valued questions; that is, sometimes, valued evaluation questions ruled out at a previous point in the process can be easily entered back in and be addressed by data being collected to answer the main questions being focused on in the evaluation. It is the evaluation team's duty to look for synergies and opportunities to give the stakeholders as much as possible in return for the time and resources they are committing to the evaluation. Finally, by this stage of the process, the evaluation team and stakeholders should strive for piercing clarity about (a) which evaluation questions will be answered, (b) how credible evidence will be gathered, (c) how the evidence will be used to justify evaluation conclusions, and (d) how the evaluation team will facilitate use of the evaluation and share the lessons learned (cf. Centers for Disease Control Program Evaluation Framework, 1999; Donaldson & Lipsey, 2006).

CONCLUSIONS

Chapter 2 and the sections just mentioned are intended to provide readers with a general overview of the program theory-driven evaluation science process adapted for effectiveness evaluations. As you should now see, it involves a highly interactive or participatory process involving evaluators and stakeholders. Unlike efficacy evaluations where the evaluators typically have much more control, determine the evaluation questions to pursue, and select the evaluation methods and design, effectiveness evaluations are highly dynamic and unpredictable. Unfortunately, there are very few examples in the literature today that illustrate how program theory-driven evaluation science unfolds in practice. Evaluators and students of evaluation are often left wanting or wondering about how this theory of evaluation practice actually looks or plays itself out in real world evaluations. Therefore, the main purpose of this book is to fill that gap in the evaluation literature.

Part II of the book focuses on applications of program theory-driven evaluation science. In the chapters that follow, you will read about actual effectiveness evaluations that were conducted using the program theory-driven evaluation approach. Please note these are presented as realistic evaluations, as opposed to exemplary evaluations, that have resulted from using this approach in practice. Each evaluation presented promises to add flesh to the somewhat bare bones of the many theoretical writings in this area, and to provide concrete examples and illustrations of what has actually happened while using this approach in real-world evaluation settings.

In chapter 4, an overview of four programs that were developed within the framework of a $20 million Work and Health Initiative (WHI) is provided. This will include descriptions of the mission of the initiative, empirical connections between work and health, and an overview of how evaluation was used to guide the strategic management of this initiative and its various program components. This introductory chapter is followed by chapters 5, 6, 7, and 8, which provide more detailed information about each program theory-driven evaluation. These chapters illustrate the results of (a) developing program impact theory, (b) formulating and prioritizing evaluation questions, and (c) answering evaluation questions. Detailed results for each evaluation are presented so that readers have complete examples and can see what kind of information and services stakeholders actually receive from participating in this process. Chapter 9 discusses crosscutting findings and lessons learned about the WHI across the four evaluations. Chapters 10 and 11 conclude part II of the book—the applications section—by presenting a new case to be solved by evaluation science. This final application focuses on an educational program in distress, and illustrates how to use program theory-driven evaluation to develop an evaluation proposal with stakeholders to help this particular program.

Part III of this volume is intended to help clarify what we now know about program theory-driven evaluation science from applications in contemporary practice. Chapter 12 provides reflections and synthesizes lessons learned. Finally, chapter 13 ends the book by exploring possibilities for the future of evaluation science.

PART

II

APPLICATIONS OF PROGRAM THEORY-DRIVEN EVALUATION SCIENCE

4

The Evaluation of a $20 Million, Multiprogram, Work and Health Initiative

The purpose of this chapter is to provide the background and rationale for The California Wellness Foundation's (TCWF) Work and Health Initiative (WHI), and to provide an overview of how program theory-driven evaluation science was used to design the initiative evaluation. The chapter begins with a brief summary of the literature that justified TCWF's decision to invest $20 million dollars to promote health through work in the State of California (Donaldson, Gooler, & Weiss, 1998), and a discussion of the specifics of the mission of the WHI. Next, the evaluation approach and WHI framework are provided to illustrate the roles evaluation played in this rather complex project. This background chapter is designed to provide readers with the broader context in which the subsequent four program theory-driven evaluations (chap. 5 through chap. 8) were embedded.

PROMOTING HEALTH THROUGH WORK

Most adults spend the majority of their waking hours engaged in work-related activities. If the notion of work is broadened to include children and adolescents pursuing education goals (often in direct pursuit of paid employment), child rearing, homemaking, and volunteerism, work may be the central activity of modern life. In fact, Freud (1930) argued that "work had a more powerful effect than any other aspect of human life to bind a person to reality," and that "the life of human beings has a twofold foundation: The compulsion to work, created by external necessity, and the power of love" (as cited in Quick, Murphy, Hurrell, & Orman, 1992, p. 3).

The links between work and health, and the potential of promoting health in the work environment has received much attention in recent years (Brousseau & Yen, 2001; Donaldson & Bligh, 2006; Donaldson et al., 1998).

During the 1980s and 1990s, the notion of the healthy worker and promoting health at work became a popular focus for researchers and organizations alike. For example, by the early 1990s, national survey data showed that 81% of companies with 50 or more employees offered at least one health promotion activity (Department of Health and Human Services [DHHS], 1993). Comprehensive worksite health promotion (WHP) programs were commonplace, and typically consisted of diagnostic or screening, education, and behavior-change activities initiated, endorsed, and supported by an employing organization and designed to support the attainment and maintenance of employee well-being (Donaldson, 1995a). Although some researchers found positive health effects and organizational benefits from WHP (e.g., Pelletier, 1993), enthusiasm for WHP programs seemed to subside by the turn of the century. Controversies over evidence attempting to substantiate the "bottom line" organizational benefits of WHP coupled with difficulties in motivating employees to participate and change their lifestyles, are at least two of the primary reasons why the approach of promoting "health at work" seemed to lose momentum (Donaldson et al., 1998).

TCWF's WHI represented a bold innovation in WHP. Instead of viewing the worksite as a setting to deliver traditional health promotion services ("promoting health at work"), WHI attempted to "promote health through work" (cf. Donaldson et al., 1998). The promise of this approach rested on the fact that much research had shown that work is of primary importance both socially and personally for individuals throughout the world. For example, work (a) contributes to one's economic well-being, (b) establishes patterns of social interaction, (c) imposes a schedule on peoples' lives, and (d) provides them with structure, a sense of identity, and self-esteem (Donaldson & Bligh, 2006; Donaldson & Weiss, 1998). Work also provides other people with a means of judging status, skills, and personal worth. Therefore, the nature of one's work (e.g., the presence or absence of work; the conditions of work) is likely to be a major determinant of his or her health status, well-being, and overall quality of life (cf. Dooley, Fielding, & Levi, 1996; Karasek & Theorell, 1990).

Brousseau and Yen (2001) mapped out some of the most central connections between work and health in modern times. Their review of the extant research literature revealed 12 work and health themes, and showed that work and health are believed to be connected through both direct and indirect causal relationships (see Table 4.1). First, there is a wealth of evidence showing that work is central to social status, which is one of the most powerful predictors of health outcomes (Adler, Marmot, McEwen, & Stewart, 1999). Some research even suggests that

TABLE 4.1
Connections Between Work and Health

- Work is central to social status, one of the most powerful predictors of health outcomes.
- Unemployment is associated with large numbers of health risks.
- Inadequate employment is also associated with poor health outcomes.
- The degree of control that employees exercise over their work influences health.
- At least one study indicates that every step up the occupational ladder has positive health consequences.
- Access to health insurance comes primarily through the workplace and has important health consequences.
- Worksite health promotion programs improve the health of those who have access to them.
- Despite its overall health benefits, work can be hazardous to health.
- Work influences the health of families and children.
- Health conditions affect work status.
- Income inequality affects health.
- An emerging social science framework integrates market conditions, employment experiences, and health.

Note: From *Reflections and connections between work and health* (pp. 3–8) by R. Brousseau and I. Yen, 2000, Thousand Oaks: California Wellness Foundation. Copyright © 2000 by California Wellness Foundation. Adapted with permission.

every step up the occupational ladder has positive health outcomes (Marmot, 1998). Unemployment, underemployment, and the degree of control that employees exercise over their work have been clearly linked to health risks and poor health outcomes (Dooley et al., 1996; Dooley & Prause, 1998; Jin, Shah, & Svoboda, 1999). Finally, access to health insurance and worksite health promotion programs comes primarily through employers, and has positive health consequences (Schauffler & Brown, 2000).

It is also well documented that work conditions in the American workplace are undergoing fundamental redefinition (Adkins, 1999; Donaldson & Weiss, 1998; Howard, 1995). Advances in computer and information technologies, intensified global competition, increasing workforce diversity, corporate downsizing and reengineering, and the movement away from manufacturing toward service and information occupations and economies are having a profound effect on the way employees work and live. For example, the traditional psychological contract between employers and employees was based on trust and the exchange of employee loyalty and effort in return for wages, security, and reasonable working conditions. The new psychological contract is often characterized by self-interest, career commitment (rather than organizational commitment), and mistrust on the part of employees (Rousseau, 1996). Furthermore, a decade of corporate restructuring and downsizing has made the temporary employment business, where employees typically work for limited periods without benefits (including health care and health promotion), the single largest job-creator industry (Aley, 1995). There is, and will continue to be, a shift toward more "contingency work," frequent turnover or job churning, sequential careers (changing professions), and jobs requiring new and different skills (Donaldson & Weiss, 1998; London & Greller, 1991).

It was believed that California workers had been particularly hard hit by some of these trends (Donaldson et al., 1998; Brousseau & Yen, 2001). For example, large-scale restructuring and downsizing appeared to have left many California workers with reduced wages, unemployed, underemployed, stuck with temporary work, without adequate benefit packages, or generally with low quality work arrangements. Given trends toward an increasingly diverse California workforce, it was believed that these problems were most severe for minorities, women, younger workers, and those without higher education.

These trends inspired the Board of TCWF to invest $20 million in a 5-year, statewide Work and Health Initiative. The WHI aspired to promote the health and well-being of California workers and their families by improving the nature and conditions of work in California. Simply put, this initiative rested on the notion that "helping Californians find and succeed in good jobs may be California's most promising health promotion strategy" (Donaldson, Gooler, & Weiss, 1998, p. 171). A general description is now provided of the WHI and of how program theory-driven evaluation science was used to develop the various programs and organizations charged to "promote health through work" throughout the State of California.

MISSION OF WORK AND HEALTH INITIATIVE

The mission of the WHI of TCWF was to improve the health of Californians by funding employment-related interventions that would positively influence health. Fundamental to this initiative was the perspective that relationships between work and health were shaped by an evolving California economy. The goals of the initiative were (a) to understand the rapidly changing nature of work and its effects on the health of Californians; (b) to increase access to high quality employment for all Californians; (c) to improve conditions of work for employed Californians; and (d) to expand the availability of worksite health programs and benefits. The means of achieving these goals were:

- To better understand the links between economic conditions, work and health and their effects on employees, employers, and the community.

- To demonstrate the relationship(s) between work and health for employees, potential employees, employers, and the larger community.
- To influence corporate and government policymakers to improve working conditions for individuals in California.
- To establish a self-sustaining network of researchers and practitioners to advance the work and health agenda.
- To evaluate the process and impact of all TCWF funded projects aimed at improving the work and health of Californians.

To accomplish these objectives, TCWF funded four programs composed of over 40 partner organizations working together to improve the well-being of Californians through approaches related to employment. The Future of Work and Health and the Health Insurance Policy programs were expansive and comprehensive research programs designed to generate and disseminate knowledge of how the nature of work was being transformed and how that change will affect the health and well-being of Californians. In the Health Insurance Policy program, current statewide trends related to health and health insurance within California were studied through extensive survey research on an annual basis. In the Future of Work and Health program, researchers throughout California examined the changing nature of work and health, and identified some implications for improving working conditions and lowering employment risks.

The initiative also included two demonstration programs in 17 sites throughout the state to assist both youth and adults in building job skills and finding employment. The Winning New Jobs program aimed to help workers regain employment lost due to downsizing, reengineering, and other factors driving rather dramatic changes in the California workplace, and thereby put an end to the adverse health consequences that most workers experience as a result of unemployment. Finally, the Computers in Our Future program aimed to enable youth and young adults from low-income communities to learn computer skills to improve their education and employment opportunities, thereby improving their own future health as well as the health and well-being of their families and communities.

EVALUATION APPROACH AND INITIATIVE FRAMEWORK

Systematic program theory-driven evaluation was used to guide the strategic management of each program in the initiative, as well as to inform the entire initiative. The initiative evaluation team served as an integrating, synthesizing force in evaluating goals, objectives, strategies, and outcomes central to the long-term impact of the initiative. Crosscutting goals and synergies were identified, enhanced, and evaluated in an effort to maximize the overall impact of the initiative. In addition, the initiative evaluation team developed evaluation systems that provided responsive evaluation data for each program. Those data were used to continually improve program effectiveness as well as evaluate long-term outcomes. Figure 4.1 illustrates the initial conceptual framework that was developed by the evaluation team in collaboration with relevant stakeholders to summarize this complex effort. It is important to note, however, that the overall framework and specific program goals evolved during the implementation period of the initiative. Whereas there was initial emphasis on maximizing synergy among grantees, this was deemphasized as an important goal by TCWF after the first year of the initiative.

To ensure that the perspectives and problem-solving needs of those with a vested interest in the initiative programs (e.g., TCWF, grantees, program administrators, staff, and program recipients), collectively known as stakeholders, were understood and addressed, the evaluation team adopted a participatory program theory-driven evaluation science approach.

Key objectives of this approach were to empower stakeholders to be successful, facilitate continuous program learning, assist with ongoing problem-solving efforts, and to facilitate improvement at as many levels as possible throughout the life of the initiative. Decisions about evaluation design, goal setting, data collection, program monitoring, data analysis, report development, and dissemination were highly collaborative.

The program theory-driven evaluation approach rested on developing program impact theories for each program and using evaluation data to guide program development and implementation. Each program impact theory was developed collaboratively and was based on the stakeholders' views and experiences, prior evaluation and research findings, and more general theoretical and empirical work related to the phenomena under investigation. Such frameworks provided a guiding model around which evaluation designs were developed to specifically answer key evaluation questions as rigorously as possible given the practical constraints of the evaluation context.

Data Collection

Data collection efforts were based on the premise that no single data source is bias free or a completely accurate representation of reality. Evaluation plans were designed to specifically encourage each grantee

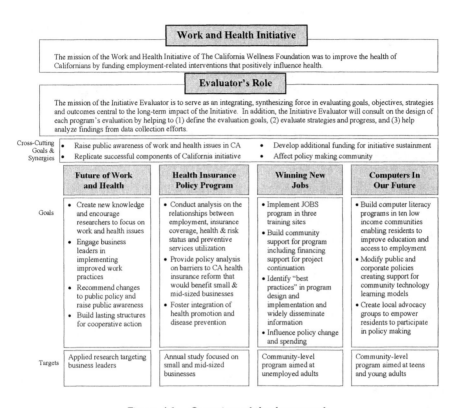

Figure 4.1. Overview of the framework.

to utilize multiple data collection strategies with different strengths and weaknesses (Chen, 1997; Cook, 1985; Donaldson, 2003; Shadish, 1993). A special effort was made to understand cultural and language concerns so that the methodologies employed yielded accurate data. In addition to evaluating program outcomes and impacts, evaluative efforts were both formative (i.e., aimed at developing and improving programs from an early stage) and process-oriented (i.e., geared toward understanding how a program achieves what it does over time). In subsequent chapters, a range of quantitative and qualitative data collection efforts for each WHI program will be described in more detail.

FORMATIVE EVALUATION TOOLS

To support continuous program improvement throughout the life of the initiative, the evaluation team:

- Provided midyear evaluation reports.
- Facilitated midyear conference calls to discuss program evaluation findings and recommendations with grantees and TCWF program officers.
- Provided year-end evaluation reports.
- Facilitated year-end conference calls to discuss program evaluation findings and recommendations with grantees and TCWF program officers.
- Provided grantees an opportunity to evaluate TCWF program officers and the evaluation team on an annual basis.

In addition, these efforts were supplemented with several interim evaluation reports and frequent communications with grantees and TCWF program officers to provide timely feedback based on evaluation data collected throughout the year.

Summative Evaluation

The evaluation team collected and analyzed extensive quantitative and qualitative data pertaining to the impact of the WHI. Approximately 200 evaluation reports were written and provided to grantees and/or TCWF throughout the life of the initiative. In an effort to determine the most useful format and content for the summative evaluation reports, the evaluation team initiated several discussions with the leadership of TCWF. As a result of the discussions, the evaluation team provided a final summative evaluation report to the board of TCWF that conformed to the following guidelines:

- The main purpose of the report was to provide a summary of evaluation findings and conclusions in a relatively brief manner.
- Qualitative as well as quantitative findings were presented.
- The report reflected the evaluation team's candid evaluation of the WHI from an external evaluation perspective, and did not necessarily reflect the views of the grantees or the foundation staff involved with the project.
- The final summative evaluation report was a confidential internal document.

On request, the evaluation team offered to provide copies of supporting documents, previous evaluation reports, data tables, or to conduct additional data analyses to justify or expand on findings and conclusions presented in the final summative evaluation report. The evaluation team also provided distinct summative evaluation reports for each program, each site within a program, and produced and disseminated public documents describing key findings and lessons learned from the WHI.

CONCLUSION

In the following chapters, you will learn about the application of program theory-driven evaluation science to each of the four programs that made up TCWF's WHI. A particular focus of these examples will be to explore how program theory-driven evaluation science was used to develop the programs and organizations funded by TCWF to implement the general design of the WHI described in this chapter. A common organizing framework is used across the four programs which consists of (a) developing program impact theory, (b) formulating and prioritizing evaluation questions, (c) answering evaluation questions, and (d) the benefits and challenges of using program theory-driven evaluation science for the specific case. In addition, crosscutting WHI findings and lessons learned, and additional lessons from the practice of program theory-driven evaluation science across a range of program domains, is discussed in chapter 12. Finally, future directions for using evaluation science to develop people and organizations will be explored in chapter 13.

5

Evaluation of the Winning New Jobs Program

The purpose of this chapter is to illustrate how program theory-driven evaluation science was used to help design, implement, and evaluate a reemployment program delivered in three different organizations in the state of California. This chapter is organized to illustrate how the three general steps of program theory-driven evaluation science were implemented in this evaluation:

1. Developing program impact theory.
2. Formulating and prioritizing evaluation questions.
3. Answering evaluation questions.

The following discussion of the evaluation of Winning New Jobs (WNJ) is intended to provide a specific example of program theory-driven evaluation in practice, and to demonstrate the benefits and challenges of using this evaluation approach with a rather diverse set of stakeholders working in very different geographic regions and settings.

DEVELOPING A PROGRAM IMPACT THEORY FOR WNJ

Program Description and Rationale

The original mission of the WNJ program was to provide job search training to 10,000 unemployed and underemployed Californians over a 4-year funding period. This project was based on a theory-based intervention, JOBS, which was developed and initially tested via a randomized controlled trial (RCT) in Michigan by the Michigan Prevention Research

Center (MPRC; Vinokur, van Ryn, Gramlich, & Price, 1991). To accomplish these goals, under the sponsorship of TCWF, Manpower Demonstration Research Corporation (MDRC) subcontracted with MPRC and, together with TCWF, used systematic organizational readiness assessments to select and train staff at organizations in three diverse California communities (see Donaldson et al., 1998): Los Angeles County Office of Education, NOVA Private Industry Council, and Proteus, Inc. A brief description of each WNJ implementation site is now provided.

Los Angeles County Office of Education.   The organization selected to implement WNJ in southern California was the Los Angeles County Office of Education (LACOE). LACOE provided a broad range of education and employment services throughout Los Angeles County. For example, at the time LACOE was selected for WNJ, it was providing job-search training to over 20,000 participants a year through its regional training centers and was running the largest job-training program for high school students and adults in California. LACOE offered the WNJ program at its regional sites located throughout Los Angeles County.

NOVA Private Industry Council.   The second organization selected was the NOVA Private Industry Council in northern California. At the time of selection, NOVA had been providing employment and training services to people in Santa Clara and Alameda Counties since 1983 and was serving over 9,000 individuals every year. NOVA delivered the WNJ intervention at the Silicon Valley Workforce Development Center, a "one stop" center that had been nationally recognized for its state-of-the-art training programs.

Proteus, Inc.   Proteus, Inc. was the third organization selected to deliver the WNJ program in central California. Proteus was a community-based, nonprofit organization that had been providing adult education and job training for over 23 years in the Central Valley. With WNJ, Proteus focused on serving unemployed people in Fresno County through three of their county offices.

Program Impact Theory Development

Prior to site selection, the evaluation team began working with the initial grantees, MPRC and MDRC, and staff from the TCWF to develop a program impact theory of how participation in the WNJ program would lead to reemployment and promote the mental health of California workers from the diverse regions and communities. Once the sites were funded, staff from each organization joined the discussions about

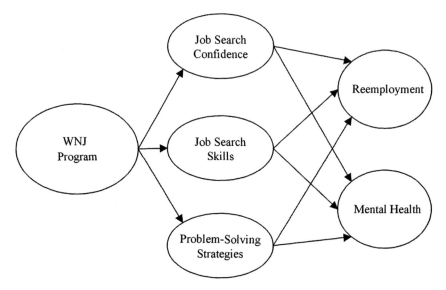

Figure 5.1. WNJ program impact theory.

program impact theory. These discussions occurred at various face-to-face meetings, during conference calls, and over e-mail.

Although these discussions continued throughout the duration of the evaluation, a parsimonious version of the program theory was agreed upon by these stakeholders and used to guide the evaluation across the three sites.

The core program impact theory used to guide the evaluation of WNJ is shown in Figure 5.1. Participants attended five half-day workshops over a 1-week period. The workshops were designed to improve job search self-confidence, job-search skills, and problem-solving strategies, including inoculation against setbacks (i.e., expectations of setbacks). These skills and psychological factors were presumed to facilitate reemployment and improve mental health. Furthermore, the WNJ program was hypothesized to have impacts at multiple levels: participant (e.g., increased job search, self-efficacy, and reemployment), organization (e.g., staff skill development, reputation enhancement), community (e.g., increased access to job search services), and the policy environment (e.g., financial support for the continuation of the program).

This conceptualization of the WNJ program was used to develop and prioritize evaluation questions that were agreed upon by program stakeholders. More specifically, during the earliest phase of program implementation, leaders from each of the three WNJ sites, as well as the program management team (MDRC, MPRC, and TCWF) and the evaluation team, agreed on five core areas for the evaluation to address.[1] Evaluation resources would subsequently be implemented to answer core questions in these five areas:

- Program implementation.
- Program service.
- Short-term outcomes.
- Reemployment outcomes.
- Program sustainability and replication.

FORMULATING AND PRIORITIZING EVALUATION QUESTIONS

In much the same way that the discussion of program theory was participatory and highly interactive, the discussion of potential evaluation questions was a collaborative process including the relevant stakeholders. However, it was much more difficult to gain consensus around prioritizing evaluation questions than it was to develop a program impact theory for WNJ. The focal point of the conflict and concerns about the evaluation questions revolved around the strong desire of some stakeholders to make determining program impact with a high degree of certainty, the highest priority for the evaluation. Other stakeholders felt that it would be nearly impossible to implement a randomized controlled trial with a high degree of fidelity in the WNJ organizations, and that effort to do so would not provide unequivocal results. After much conflict and discussion, TCWF sided with the coalition opposed to conducting a randomized controlled trial, and encouraged the stakeholders and evaluation team to focus on other types of evaluation questions.

Another series of discussions with relevant stakeholders attempted to develop agreement about the evaluation questions in light of TCWF's decision to discourage a randomized controlled trial for the WNJ evaluation. After exploring a wide range of possible evaluation questions and considering resource constraints and the limits of each implementing organization, the relevant stakeholders decided to focus the WNJ evaluation on the following five areas:

---

[1]The process of identifying and prioritizing areas the evaluation addressed resulted in program stakeholders agreeing that the evaluation would not focus on some of the initial hypothesized relationships noted here (e.g., links to mental health outcomes).

1. *Program implementation.* Can the Michigan JOBS program be implemented in different types of service organizations in California? What does implementation look like? What are the key challenges and success factors to implementation?

2. *Program service.* Whom are the sites serving (e.g., population characteristics of service recipients)? How many people are served at each site?

3. *Short-term outcomes.* Does WNJ increase people's confidence in their ability to use their newly acquired and/or enhanced job-seeking skills?

4. *Reemployment outcomes.* Do people find employment? And, what does their employment situation look like?

5. *Program sustainability and replication.* Does WNJ generate resources for program sustainability beyond the life of the grant? Do other California organizations learn about and adopt the WNJ model?

ANSWERING EVALUATION QUESTIONS

This section provides a rather detailed account of the third step in the program theory-driven evaluation science process. To help readers understand how Step 3 can unfold in practice, data collection, key evaluation findings and considerations for future programming are presented. These topics illustrate how one transitions from the conceptual Step 1 and Step 2 to the empirical Step 3 in the program theory-driven evaluation process. This section is particularly important in light of the recognized need for detailed examples illustrating how program theory-driven evaluation science has been implemented in recent evaluation practice.

Data Collection

A range of evaluation designs and data collection methods for answering the five questions were explored in detail with the stakeholders. Although consensus was never reached, the vast majority of the stakeholders conceded that it would not be advisable to conduct a randomized controlled trial (like prior evaluation research on JOBS) or to implement an evaluation of a meaningful control or comparison group. Instead, a variety of quantitative and qualitative data about the WNJ program were collected to provide data that would allow drawing evaluative conclusions about the five core areas, at least in light of design limitations and consideration of alternative explanations.

For example, extensive standardized eligibility, demographic, pretest, posttest, and employment follow-up data were collected at each site. Overall, data were collected for over 5,100 individuals, including eligibility, demographic, and pretest data for 4,960 individuals, posttest data for 3,684 individuals, and employment follow-up forms for 3,476 individuals who completed the workshop. These response rates were considered adequate for the nature of the program and types of analyses conducted. In addition to these data, various types of qualitative implementation and outcome data were also collected. Further, databases tracking participants in other parts of the country and world were available for comparison purposes. This collection of databases was used for both the formative and summative evaluation of the WNJ program.

To support continuous program improvement within the WNJ implementation sites throughout the life of the initiative, the evaluation team prepared and disseminated 44 evaluation reports to WNJ program leaders and program managers over the 4-year funding period. These included four year-end evaluation reports, four midyear evaluation reports, 29 site reports, and seven interim reports to program grantees. As tools for program monitoring and improvement, these reports documented not only key accomplishments and program activities, but also key program challenges and recommendations for addressing challenges. Conference calls and/or face-to-face meetings were held to discuss each report. During these communications, the evaluation team facilitated discussion of the findings and strategies to addressing program recommendations. Although early challenges and recommendations focused on program development, recruitment, service delivery and data collection efforts, later challenges and recommendations centered on developing and finalizing program institutionalization and dissemination plans.[2]

Key Evaluation Findings and Conclusions

One area that generated much discussion throughout the continuous improvement process was WNJ service goals. The initial program goal of serving 10,000 Californians, as stated in the original Request for Proposals, was lowered to 6,500 individuals prior to site selection. This reduced number represented program leaders' estimates based on past experience and estimates of future needs. Having a reliable tracking system in place enabled the evaluation team to send early warning signs that sites were having a difficult time reaching the target population of job

---

[2]Detailed discussions of program challenges and recommendations for addressing challenges were provided in midyear and year-end evaluation reports.

seekers. Further discussions revealed the original program eligibility requirements (e.g., age, length of unemployment, employment history, and depression) were more difficult to meet than expected. In an effort to overcome this recruitment barrier, eligibility requirements were relaxed (with the exception of screening for depression) in July 1998 and monitored by the evaluation team.

Economic conditions were another factor that appeared to affect recruitment. At the start of the initiative, unemployment rates in the three service delivery areas were between 2% and 4.7% higher than they were at the initiative's conclusion.[3]

As a result of the decreasing numbers of people looking for jobs and the increasing challenge of recruiting clients, sites began discussing these issues on a bimonthly basis with their program management team. On a regular basis, site leaders began to discuss and share effective recruitment practices. Despite these efforts to increase participation in WNJ, the WNJ program management team, including TCWF, agreed during the first half of 1999 to lower the overall program service goal to 5,000.

Summary Answer to Evaluation Question Number 1—Program Implementation

The Michigan JOBS program was successfully implemented in three diverse service organizations in California as WNJ. Serving diverse populations in different geographical regions, these organizations implemented the WNJ curriculum with only minor curriculum adaptations over time. Participants felt that the behaviors of facilitators and the group dynamics during training sessions were consistent with those targeted in the WNJ model.

Supporting Evidence for Answering Evaluation Question Number 1. During the start-up phase of the program (first half of 1997), each of the three WNJ organizations (LACOE, NOVA, and Proteus) participated in two training sessions to learn about the WNJ model and develop skills in delivering the curriculum. The first training was held in Michigan over 2 weeks in February and included both program managers and facilitators. The second training session focused on facilitators and took place in May during a 1-week period in Oakland, California. Following these initial trainings, each site became fully operational and started offering workshops to job seekers by July 1997.

---

[3]Data from the Employment Development Department show that from March 1996 to September 2000, unemployment rates lowered in each program service delivery area: from 16.2% to 11.3% in Fresno; from 8.2% to 5.5% in Los Angeles; and from 3.7% to 1.7% in Santa Clara.

As of December 2000 (the final month of data collection), the three WNJ sites conducted a total of 455 workshops, including 207 at Proteus, 119 at NOVA, and 129 at LACOE. The workshop consisted of 5 days of training for a total of 20 hours. From 2 to 40 participants attended the workshops, with an average number of 14 participants per workshop. Overall, 88% of all participants completed the workshop (i.e., attended at least four of the five training sessions). This completion rate is very favorable for a voluntary job-search training program. Among those who did not complete at least 4 of the 5 days of the workshop, 27% ($n = 173$) were either already in new jobs or were interviewing, which prevented them from attending.

Evidence that the program was implemented as intended came from both observational findings and posttest survey results. Structured observation forms were used by program managers over the course of the project and suggested that WNJ facilitators adhered to the WNJ curriculum content very closely. In addition, self-report data were collected at the completion of each workshop to assess participants' perceptions of the workshop (i.e., trainer behaviors, social attractiveness of the group—as targeted in the WNJ model). Posttest feedback from participants showed that nearly all (92.2%) of WNJ clients who completed the posttest reported that the trainers had engaged in the expected, supportive behaviors "pretty much" or "a great deal" of the time, and more than four fifths (85.7%) rated the trainers "somewhat" or "very" positively. Similarly, more than four fifths (82.2%) of WNJ clients who completed the posttest reported that positive group dynamics occurred "pretty much" or "a great deal" of the time, and rated the group "somewhat" or "very" positively (87.3%).

With respect to the program itself, it was well liked by both participants and trainers. Feedback from participants revealed several factors they felt made the workshop particularly effective. These included: (a) the positive, supportive, and motivating nature of the workshops; (b) helpfulness of the interview activities; (c) interacting with a group of people; (d) skits and role-playing exercises; and (e) program training exercises. Feedback from program staff revealed that it was not uncommon for participants to organize potlucks at the end of the workshops and to maintain contact with one another to support each other's job-search efforts.

Although participants gave fewer suggestions for improvement, these tended to concern the length of the sessions, needing more breaks, and wanting more clarification for specific group exercises and activities.

Among facilitators, one striking finding was that WNJ trainers developed a genuine affinity for the program. As noted by one trainer, "It takes the burden off me and allows me to draw upon the expertise and experience of participants." As noted by another facilitator, "I've been facilitating these kinds of training sessions for years. I thought I knew everything. This program gave me a chance to develop new skills—I now am so much more energized and excited when I'm out there." As reflected in these comments, many facilitators frequently commented on how much they liked the "spirit" of the training method (i.e., positive and supportive, lacking negative feedback) and that it acknowledged that there is "no one best way" to approach job searches and that everyone has strengths and skills to draw on. Another factor that made the WNJ program well liked among facilitators entailed the opportunity to facilitate sessions with other trainers. As one trainer noted, "I don't feel exhausted like I used to. I can switch being 'on' with my partner and share the responsibility of facilitation." Another noted, "I get a lot of great ideas for improving my own performance by learning from those I train with."

Several factors appeared to contribute to the successful implementation of the JOBS model, including: (a) extensive training of program staff in the WNJ model and training protocol; (b) strong involvement of the program management team (MDRC, MPRC); (c) active monitoring of deviations from recruiting and training protocols by the sites and program management team; (d) refresher training sessions; and (e) continuous improvement feedback from the evaluation team. In addition, the program began with a well-developed model and training protocol. Implementation efforts did not need to focus on developing a program curriculum. Instead, it required building an understanding of the model and building new types of facilitation skills among program managers and facilitators. Although there was support for piloting the program with different populations (e.g., welfare-to-work, work furlough populations), it appeared that the program management team was less willing to accept deviations in the training protocol itself. That is, efforts to change content or offer only part of the curriculum in fewer sessions were discouraged. Consistent with this, facilitators participated in bimonthly conference calls with WNJ program developers (MPRC), during which time they discussed such things as tactics for positive reinforcement and methods of handling difficult clients. This provided an opportunity to discuss operational challenges and share ideas and approaches for addressing those challenges.

Factors that made program implementation challenging concerned the nature of the program itself and gaining top management buy-in. More specifically, key challenges included (a) the staff-intensive nature of delivering WNJ (protocol requires two staff per session); (b) time investment—the WNJ protocol took 20 hours and 5 days to complete; (c) eligibility requirements made it difficult to reach target service goals; and (d) job seekers as well as program leaders had a difficult time

TABLE 5.1
Demographic Characteristics of WNJ Population Served

	OVERALL (N = 4,774)	LACOE (N = 1,744)	NOVA (N = 1,151)	PROTEUS (N = 1,879)
Gender				
Male	40.0%	35.7%	43.6%	41.7%
Female	60.0%	64.3%	56.3%	58.3%
Average Age	38.0 years	41.7 years	44.8 years	30.9 years
Race/Ethnicity				
Latino/a	37.8%	39.5%	12.1%	51.3%
African American	13.1%	16.1%	4.8%	15.3%
Asian American	10.6%	11.5%	20.2%	4.1%
European American	31.3%	27.2%	56.0%	20.6%
Mixed/Other	7.1%	5.6%	7.0%	8.6%
Education				
Less than HS	15.4%	9.4%	3.2%	29.3%
HS/GED	29.4%	28.3%	14.7%	40.1%
Some college +	55.2%	62.3%	82.1%	30.6%

distinguishing what was unique about the WNJ program prior to participating in, or observing, the workshops.

Summary Answer to Evaluation Question Number 2—Program Service

The three WNJ sites served 5,290 unemployed or underemployed Californians in 455 WNJ workshops over a 4-year period. WNJ program sites served a very diverse population of Californians in terms of ethnicity, age, educational background, and employment history. Nearly two thirds (60%) of WNJ participants were female; more than two thirds (68.7%) were people of color; more than two fifths were over 40 years of age (41%); and nearly one half held a high school degree or had fewer years of formal education (44.8%).

Supporting Evidence for Answering Evaluation Question Number 2.
As shown in Table 5.1, each WNJ site served unique and diverse populations. Whereas participants at NOVA tended to be the oldest and most educated group overall, those at Proteus tended to be younger and the least likely to have an education beyond a high school degree. All sites served more females than males.

TABLE 5.2
Employment Histories of WNJ Population Served

	OVERALL	LACOE	NOVA	PROTEUS
Job Tenure	4.2 years	5.3 years	5.8 years	2.2 years
Salary ($ per hour)	$13.22	$13.59	$21.52	$7.91
Had Health Benefits	51.7%	56.9%	72.1%	34.1%
Length of Unemployment	12.9 months	16.3 months	11 months	11.2 months
% with 2+ jobs in past 3 years	41.0%	34.9%	39.0%	46.8%

The employment history table (Table 5.2) shows that participants were in their last job on average for 4.2 years, worked 8.2 hours per day, and earned an average of $13.22 per hour. Two fifths (41%) of all WNJ participants held two or more jobs during the past 3 years. Slightly less than half did not receive benefits in their last job. With regard to unemployment status and length of unemployment, the data indicate that WNJ clients report a much higher length of unemployment at time of participating in WNJ compared to the state average (12.9 months vs. 12.2 weeks).[4]

Identifying populations reached versus those originally targeted was based primarily on eligibility screening and demographic survey questions. As the intervention was designed, to be considered eligible, participants had to be adults (at least 18 years of age), seeking a job, worked for pay for at least 6 months total, left or lost a job within the last year, and not exhibiting symptoms of severe depression. According to past research, the most important eligibility screening criteria is evidence of severe depression. This was assessed with a 10 item, 5-point scale (Hopkins Symptom Checklist). Individuals were deemed ineligible if they scored above an average of 3.

Forty Percent Failed to Meet One or More of the Original Eligibility Criteria.   Overall, 36.9% (*n* = 1,831) failed to meet at least one of the eligibility criteria, while 7.8% (*n* = 386) failed to meet two or more of the eligibility criteria: 213 from LACOE, 43 from NOVA, and 120 from Proteus. Collectively, 2.4% did not meet the depression criteria, 24.5% had not lost or left a job during the previous year, 9.6% had not worked for pay for 6 months, 3.2% were not seeking a job, and 1.1% were under 18 years of age.

Initial feedback from sites indicated that WNJ may be beneficial to some individuals who do not meet one or more of the eligibility criteria. During 1998, there was a joint decision made among TCWF, MDRC, MPRC, the evaluation team, and the three site program managers to

---

[4]Source: *Unemployed persons by state, and duration of unemployment, July, 1999* (based on CPS). Bureau of Labor Statistics, 1999, Table 4D.

relax the eligibility screening criteria cap (except for depression). The evaluation team agreed to monitor the implications of relaxing these eligibility criteria on program effects, the results of which are now noted.

In order to investigate whether or not there were differences in targeted short-term (i.e., psychological) factors or employment outcomes for individuals who participated in WNJ who did not meet one or more of the eligibility criteria, a number of comparative analyses were conducted. Among those not meeting specific criteria, those who were exhibiting symptoms of clinical depression had lower pretest levels of job search self-efficacy and confidence, as well as lower posttest levels of self-mastery. They also had higher pretest and posttest expectations of setbacks than those who met the depression criteria. However, the depressed clients experienced greater increases in job search self-efficacy and self-mastery. They also experienced a decrease in expectations of setbacks, whereas those who met the criteria experienced an increase in expectations (which is hypothesized in the model to lead to better coping).

Similar comparative analyses indicated that those who had not worked for 6 months for pay and those who had not left or lost a job in the past year were less likely to benefit from the program. For example, those with longer employment experiences had lower pretest and posttest levels of self-efficacy, self-mastery, and confidence. They also rated the trainers and the group dynamics less favorably, and were less likely to become reemployed. Similarly, participants who had not worked at least 6 months for pay had lower pretest levels of job search self-efficacy and self-mastery, but higher posttest expectations of setbacks. They also rated the trainers and group dynamic less favorably, and were less likely to become reemployed than those who had worked for at least 6 months for pay. It should be kept in mind that the sample size was small for some comparisons (e.g., number of individuals meeting and not meeting the depression criteria), and thus, caution is warranted when interpreting these results.

Summary Answer to Evaluation Question Number 3—Short-Term Outcomes

The preponderance of the available evidence indicates that WNJ participants experienced significant improvements in job search self-efficacy, self-mastery, confidence to handle setbacks, and expectations of employment setbacks. However, these findings must be interpreted in light of evaluation design limitations.

Supporting Evidence for Answering Evaluation Question Number 3.
Self-report data were collected from WNJ participants prior to and after completing the WNJ workshop. These data included established measures

of job search self-efficacy, sense of self-mastery, and inoculation against employment setbacks. Analyses revealed that the findings were consistent with the WNJ program impact theory. That is, at each WNJ site, there were statistically significant improvements in participant self-efficacy, self-mastery, confidence to handle setbacks, and expectations of employment setbacks. These findings were consistent with the evaluation team interview and observational data, as well as with the published literature on the JOBS program. Additional analyses also revealed that participants who rated the group dynamics of the WNJ workshop more favorably were more likely to report higher levels of these psychological factors than those who rated group dynamics less favorably.

Evaluation Design Limitations.   The quantitative data were collected using a one-group pretest/posttest research design. That is, pretest scores were compared to posttest scores without a control or comparison group. After extensive discussion about the limitations of this type of design, TCWF, MDRC, MPRC, and the evaluation team agreed to employ the pretest/posttest research design and to gather interview and observational data to supplement the quantitative data. Therefore, we are not able to rule out with a high degree of confidence some potential alternative explanations for the short-term outcome findings or for the reemployment outcome analyses presented next. Nevertheless, as requested by TCWF, the evaluation team provided its candid evaluation based on the evidence compiled.

Summary Answer to Evaluation Question
Number 4—Reemployment Outcomes

> *Approximately 65% of all WNJ completers reported becoming reemployed 180 days after the workshop. Reemployment rates differed dramatically by site, ranging from 55% for LACOE to 60% at NOVA and 75% at Proteus.*

Supporting Evidence for Answering Evaluation Question Number 4. Each WNJ site conducted follow-up telephone interviews with completers at 30, 60, 90, and 180 days postworkshop throughout the funding period. There was no expectation that sites would continue to conduct follow-up beyond the funding end date (December 31, 2000). Thus, for those who participated in WNJ at any point after June 2000, the full 180-day follow-up was not conducted (e.g., September 2000 participants had a 30-day [November 2000] and 60-day [December 2000] follow-up only). Data were batched and sent to the evaluation team following the 90- and 180-day postworkshop dates. The evaluation team received employment follow-up surveys for a total of 3,476 participants conducted by the sites. A summary of reemployment outcomes for all follow-up data collected can be seen in Table 5.3.

Response Rates.   An average of 68% of all participants who completed the WNJ workshops were contacted during the follow-up. Cumulative data show that, overall, contact rates for the employment survey ranged from 65% to 70% across the time periods. After comparing characteristics of those who were (and were not) contacted during follow-up time periods, very few significant differences were detected, suggesting that individuals reached during the follow-up time periods were generally representative of all those who completed the WNJ workshops.

Reemployment Rates.   As expected, reemployment rates increased at each follow-up time period (from 36% at 30 days, 45% at 60 days, 51% at 90 days, to 65% at the 180-day follow-up). Among those who became reemployed, nearly two thirds (61.5%) obtained their jobs within 2 months of completing the workshop, and more than one half (59.8%) reported being reemployed in their chosen occupation. When asked about the role of WNJ in helping them obtain employment, the majority of participants perceived WNJ to be important in helping them become reemployed. This was especially true for those who became reemployed right away. For example, whereas 94% of those who became reemployed within 30 days said WNJ was "pretty much" or "a great deal" important in helping them find a job, just over four fifths (81%) said the same if they became reemployed between 3 and 6 months following the workshop.

Due to the research design limitations discussed previously and the lack of sound research on comparable programs in California, it is challenging to determine the merit of WNJ compared to other reemployment programs based on reemployment rates. However, it is interesting and potentially useful to view WNJ reemployment rates in the context of the rates reported in the original JOBS randomized trial. For example, 36% of WNJ participants were reemployed 30 days after the workshop, compared to 33% (experimental group) and 26% (control group) in JOBS. At the final follow-up assessment (6 months for WNJ; 4 months for JOBS), 65% were reemployed in WNJ compared to 59% (experimental group) and 51% (control group) in JOBS. In our view, these numbers are at least encouraging given that (a) 40% of the WNJ participants failed to meet one or more of the eligibility criteria; (b) WNJ participants were previously unemployed on average for more than 12 months (compared to 13 weeks for JOBS participants); and (c) WNJ participants were generally more

TABLE 5.3
WNJ Reemployment Outcomes

Days	Org.[a]	Total N	Response Rate		Employment Outcomes for Those Contacted		
			Contacted	No Contact	Employed	Of Employed, Number in Main Occupancy	Not Employed
30	FS	3476	70.2% (n=2441)	29.8% (n=1036)	35.9% (n=877)	58.5% (n=513)	64.1% (n=1563)
	L[b]	949	49.2% (n=467)	50.8% (n=482)	23.6% (n=110)	55.5% (n=61)	76.4% (n=357)
	N	1083	77.1% (n=835)	22.9% (n=248)	22.9% (n=191)	54.5% (n=104)	77.1% (n=644)
	P	1444	78.8% (n=1138)	21.2% (n=306)	50.6% (n=576)	60.2% (n=347)	49.4% (n=562)
60	FS	3412	66.8% (n=2288)	33.2% (n=1138)	44.8% (n=1024)	58.5% (n=599)	55.2% (n=1263)
	L	949	51% (n=484)	49% (n=465)	28.5% (n=138)	50% (n=69)	71.5% (n=346)
	N	1071	75.7% (n=811)	24.3% (n=260)	33.5% (n=272)	57% (n=155)	66.5% (n=539)
	P	1405	70.6% (n=992)	29.4% (n=413)	61.9% (n=614)	60.9% (n=374)	38.1% (n=378)
90	FS	3412	68% (n=2320)	32% (n=1092)	51.3% (n=1191)	59.8% (n=712)	48.7% (n=1128)
	L	949	55.8% (n=530)	44.2% (n=419)	36.2% (n=192)	52.6% (n=101)	63.8% (n=338)
	N	1057	79.6% (n=841)	20.4% (n=216)	42.8% (n=360)	59.7% (n=215)	57.2% (n=481)
	P	1405	67.5% (n=948)	32.5% (n=457)	67.4% (n=639)	61.8% (n=395)	32.6% (n=309)
180[c]	FS	2992	65.1% (n=1948)	34.9% (n=1045)	64.5% (n=1257)	62.5% (n=785)	35.5% (n=689)
	L	779	56.2% (n=438)	43.8% (n=341)	55.3% (n=242)	43.8% (n=106)	44.7% (n=196)
	N	1000	72.7% (n=727)	27.3% (n=273)	59.6% (n=433)	67.4% (n=292)	40.4% (n=294)
	P	1212	64.4% (n=781)	35.6% (n=431)	74.5% (n=582)	66.2% (n=385)	25.5% (n=199)

a: FS = Full Sample; L = LACOE; N = NOVA; and P = Proteus.

b: There were no reemployment surveys batched and returned to CGU from LACOE during the fourth quarter of 2000.

c: The 2992 individuals for whom 180 day follow-up data were collected are included in the 30/60/90 day totals.

Note. The total sample size varies across the employment follow-up time periods, because all of the follow-up time periods for participants who completed WNJ any time after June 2000 did not fall within the funding period, ending December 2000 (e.g., September 2000 participants would only have had the 30-day (11/00) and the 60-day (12/00) follow-up conducted). Individuals whose follow-up time frame fell outside the funding period were considered to have missing data (i.e., they are not included in the calculation of the response rate or reemployment rate).

diverse and believed to be less well-suited for the intervention than the carefully screened JOBS participants. However, these comparisons in combination with the reemployment rate differences between the WNJ sites underscore that demographic characteristics, occupational characteristics, and economic conditions are likely to play a larger role in reemployment than psychological factors under some circumstances.

Summary Answer to Evaluation Question Number 5—Program Sustainability and Replication

Although all three sites committed to institutionalize WNJ in whole or in part during their funding period, only one WNJ site (Proteus) continued to offer WNJ workshops as part of its standard array of services. A relatively small number of service organizations had delivered WNJ workshops to their local constituents.

Supporting Evidence for Answering Evaluation Question Number 5. NOVA and Proteus committed early on to institutionalize aspects of WNJ at their sites beyond the funding period. At Proteus, senior administrators committed to continue WNJ operations after the TCWF funding period, as part of the agency's standard array of services. In particular, Proteus planned to blend its REP (Rapid Employment Program) and WNJ workshop services for TANF (Temporary Assistance for Needy Families) applicants referred by the Fresno County Department of Social Services. Further, Proteus administrators planned to expand WNJ services to all of the organization's offices. Ultimately, local institutionalization of WNJ at Proteus was evidenced in its placement as the key job search skills and strategy component within its JOBS 2000 program. This job search program is now, and will continue to be, offered at Proteus host sites in Fresno as well as other facilities of Proteus throughout the Central Valley.

NOVA initially committed to incorporating aspects of the training protocol into their existing stream of services and programs. As NOVA completed their WNJ funding period, they perceived WNJ as one of the core services they planned to offer job seekers entering their system, as well as to integrate WNJ with their universal and core services system in the future. However, serving as the local Workforce Investment Act administrators following their WNJ funding, they were unable to continue offering WNJ as a direct service.

LACOE planned to incorporate elements of WNJ in a new job-search program called "Job Club 2000," which utilizes the cofacilitation techniques from the Michigan model, as well as "Passport to Success" materials developed in their current Job Club. Unfortunately, these plans were not realized due to discontinuation of government funding for their GAIN program.

   Dissemination Efforts.   Core dissemination efforts centered on developing documents describing the program, creating and disseminating a WNJ brochure and video, establishing a WNJ Web site, making presentations to local service providers, and generating print and media coverage. A key dissemination achievement included presenting the WNJ model at the 1999 CalWORKS conference. Instrumental in these efforts was the creation of a dissemination plan and clarification of roles and responsibilities for achieving dissemination goals.

   Through numerous conversations and meetings with WNJ program staff and the WNJ program management team, MDRC developed a three-part dissemination plan for the WNJ program, including (a) efforts to institutionalize the program at the original sites and offer the program to other agencies working with the three sites, (b) dissemination of program information to regional and state audiences, and (c) sponsoring WNJ curriculum seminars in an effort to replicate the model in other organizations. This plan was instrumental in helping to clarify dissemination goals as well as roles and responsibilities for achieving these goals. In support of WNJ dissemination plans, The California Wellness Foundation board of directors approved MDRC's request for an additional $100,000 in March 2000. Through the leadership of MDRC, these funds were used to support the many activities that took place to prepare for and replicate the WNJ program within other organizations.

   With respect to the first component, each site disseminated the WNJ model via two key mechanisms. First, presentations were made at partnering agencies, and information was disseminated about the program through local publications and public service announcements on local cable television stations. Second, each site demonstrated the model to outside observers from other organizations. Site-specific dissemination efforts also included mailings to schools and community organizations, participation in community meetings and/or forums in which they discussed the WNJ model, and partnering with other service providers. Each site also offered WNJ in a variety of settings, including one-stop career centers and adult education programs as well as in diverse agency settings in order to expand the client base and to expose other agencies to the WNJ model.

   In terms of reaching external audiences, one of the program highlights included a presentation of the WNJ program at the CalWORKS conference held in Orange County in December 1999. After being selected through a competitive proposal process, a team of representatives from each WNJ implementation site (LACOE, NOVA, and Proteus) and MDRC presented

WNJ to an audience of approximately 80 participants. The session was well attended and well received. In addition, a variety of dissemination products were produced, including a WNJ brochure, a WNJ Web page, and a WNJ video. The video, which was created by MPRC, highlighted key components of the model and included onsite coverage from each of the three WNJ sites in California. Copies of the video were disseminated to the three sites, MDRC, TCWF, and others interested in learning about the WNJ model developed by MPRC.

Program Replication.   Through a competitive training scholarship, eleven organizations were trained in the WNJ model in 2000. Several of these service agencies began to deliver WNJ workshops to their local constituents during the second half of 2000. A former JOBS affiliate also began delivering WNJ in San Diego in 1999.

Through numerous meetings and ongoing conversations and conference calls, program stakeholders first established program replication guidelines. These were used to clarify criteria for replicating the program. The core guidelines are:

- Prospective service providers must commit to operating the entire workshop curriculum in order to use the name "Winning New Jobs."
- Sponsorship and quality control and/or monitoring of program integrity at replication sites is the WNJ site's responsibility, which includes assessing organizational capability to implement WNJ, training facilitators, and providing technical assistance.
- Replication sites will be instructed as to the nature of the data collection instruments used in WNJ and strongly encouraged to use both the workshop observation forms and the depression screening instrument.
- Replication sites will be invited to participate in WNJ conventions or meetings.

In addition to these guidelines, numerous activities took place in support of program replication goals. Highlights of these achievements included: (a) the development of documents that described WNJ, (b) the development and dissemination of WNJ Training Seminar Scholarship Request for Application packets to over 700 agency staff located throughout California, (c) the hosting of open houses at each of the three WNJ sites to raise visibility of WNJ to prospective training applicants, (d) the review of and awarding of training scholarships to 11 applicant organizations, and (e) hosting two 1-week WNJ training seminars for 45 attendees. Despite these dissemination and replication efforts, the evidence suggested that few California service organizations delivered WNJ workshops to their constituents on a regular basis.

Considerations for Future Programming

This section highlights key information and/or lessons learned by the evaluation team that might improve future programming in this area. Key considerations include (a) improving workforce preparation and basic skill levels of California workers, (b) removing additional employment obstacles, (c) clarifying program sustainability and replication expectations, and (d) developing strategies to demonstrate WNJ is better than existing programs.

Workforce Preparation and Basic Skill Levels. Although the WNJ program is designed to enhance such things as an unemployed person's self-esteem, confidence, job-search skills, and self-efficacy, the lack of job-specific competencies appear to be a substantial barrier for many California job seekers. The lack of basic skills of many entry-level employees is generally a well-known challenge facing California employers. In many ways, the California working population is becoming increasingly mismatched with the needs of employers. The skills that many employers felt were lacking in a large proportion of new hires include basic arithmetic, computational skills, and English literacy. The problem was particularly acute among immigrant workers, who in some cases were illiterate in their native language as well as in English.

Removing Additional Employment Obstacles Prior to WNJ. Findings from the WNJ evaluation suggested that some key employment obstacles were difficult or impossible to address in a 5-day motivational workshop. These were, for example, skill-related obstacles (lack of skills and/or overqualified), workplace expectations (i.e., "fit" issues with respect to work schedules, salaries, skills), negative employment histories, discrimination issues, criminal records, physical appearance (e.g., tattoos), and a lack of resources such as transportation, child care, or health care. This suggested that the success of the WNJ curriculum might be dramatically enhanced by other services that removed nonmotivational obstacles prior to the 5-day workshop. Based on interview data, Proteus appeared to adopt this approach as part of its JOBS 2000 program. More specifically, in the new version of WNJ (developed after the evaluation), Proteus clients identified and addressed a variety of employment obstacles (e.g., obtaining a driver's license, building job skills) before participating in the motivational components of the program.

Clarifying Program Sustainability and Replication Expectations. In WNJ, feedback from grantees revealed that some felt pressured to disseminate and replicate the program at the same time they were struggling to obtain support from their own top management with respect to institutionalizing the WNJ program. There was also some confusion by sites with respect to roles and responsibilities for disseminating and replicating the program. Although it was clearly a grant requirement for MDRC to take the lead on these activities, this did not occur until midway into the funding period. Feedback from grantees suggested that, due to the complexity and importance of achieving these goals, these efforts would have been more successful if they were clarified and addressed early on and throughout the project.

Demonstrating WNJ Is Better Than Existing Programs. Site leaders reported that it was difficult to articulate and demonstrate the value added by WNJ in comparison to other available job placement and/or training programs. In fact, it was reported that some of the administrators in the site organizations believed WNJ was less effective than some of their existing services. This impression seemed to be based on unfair comparisons of reemployment rates and WNJ to more costly programs of much greater length. In some cases, this appeared to make it difficult to secure internal organizational commitment to make WNJ a core service, and to secure alternative funds to continue service beyond the grant period. This challenge could be addressed in future programming by developing strategies for directly examining and understanding differences between a new program and the best alternative program currently implemented at a site.

CONCLUSION

The purpose of this chapter was to illustrate in some detail how program theory-driven evaluation science was used to improve and evaluate a program designed to better the careers and lives of thousands of California workers. It also demonstrated how evaluation science was used to help develop and improve three organizations charged with providing a program to promote health through improving working conditions in the State of California. Summative evaluation findings demonstrated that the three WNJ organizations trained 5,290 unemployed or underemployed Californians over a 4-year period. The preponderance of the evaluation evidence indicated significant improvements in both short-term skill and psychological outcomes. In addition, evaluation evidence showed that approximately 65% of the participants who completed the WNJ program reported becoming reemployed within 180 days. A rather detailed account of the trials and tribulations at each step of the program theory-driven evaluation science process was provided to illuminate common practical challenges of contemporary evaluation practice. A summary of the lessons learned about the application of program theory-driven evaluation is presented in chapter 13.

6

Evaluation of the Computers in Our Future Program

This chapter examines how program theory-driven evaluation science was used to help design, implement, and evaluate an innovative program to improve health through technology skills training. The Computers in Our Future (CIOF) Program was implemented by 11 organizations throughout the state of California. To illustrate various aspects of this program theory-driven evaluation, the chapter is organized to address:

1. Developing program impact theory.
2. Formulating and prioritizing evaluation questions.
3. Answering evaluation questions.

The following presentation of the evaluation of CIOF provides another specific example of how program theory-driven evaluation science can unfold in practice.

DEVELOPING A PROGRAM IMPACT THEORY FOR CIOF

Program Description and Rationale

The second demonstration program, CIOF, created 14 community computing centers (CCCs) in 11 low-income California communities. The CCCs were designed to demonstrate creative, innovative, and culturally sensitive strategies for using computer technology to meet the economic, educational, and development needs of their local communities. The CIOF program attempted to explore and demonstrate ways in which CCCs can prepare youth and young adults ages 14 through 23 to use computers to improve their educational and employment opportunities, thereby improving the health and well-being of themselves, their families, and their communities.

Organizational readiness criteria were used to select 11 diverse California organizations from among over 400 applicants to accomplish these goals over a 4-year period (see Donaldson et al., 1998). These organizations were diverse with respect to organizational type, geographical location, and populations served. The California Wellness Foundation (TCWF) funded three organizations to facilitate program development and offer program management and technical assistance to these CCCs, including Community Partners, The Children's Partnership, and CompuMentor. Together, each site was to develop a program to address each of ten broad program goals outlined in the CIOF program model and presented in Table 6.1.

The organizations funded by TCWF to accomplish these goals are listed in Table 6.2. Collectively, these centers provided access to more than 200 computer workstations statewide. With respect to open access service goals, each site collectively committed to providing unrestricted open access to 27,705 Californians (approximately 6,900 individuals per year, statewide). Similarly, they committed to providing technology training and work experiences to over 4,300 youth and young adults over the 4-year program period. A brief description of each organization is now provided.

Break-Away Technologies is a nonprofit organization working to create access to the latest in information technology and computer training for low-income residents of Los Angeles, primarily African American and Latino youth and young adults. Break Away Technologies was established to provide underserved communities with access to cutting-edge digital technologies, high-technology solutions, software applications, future-directed training, creative technology careers, e-commerce solutions, and "new-tech community development strategies" in environments that are innovative, supportive, nurturing, and safe. Break-Away Technologies is committed to facilitating the development of "smart neighborhoods" in actual and virtual communities with dynamic connectivity that links people to people, people to services, people to resources, and people to information in language-appropriate, community-centered, creative, and progressive ways that advance personal success and community development and that are culturally sensitive. Break-Away Technologies provides telecommunication access and a wide range of training and services for children, youth, and community organizations in south central Los Angeles. It was hoped that CIOF would enable Break-Away Technologies to develop

TABLE 6.1
CIOF Program Goals

- *Create Center*: Build 11 community computing centers (CCCs), which emphasize each of the CIOF program model components.
- *Access*: Increase access to computer technology and training and offer periods of largely unrestricted use of computers to low-income residents in the local community who would otherwise lack access to computer technology.
- *Education and Skills Training*: Develop a raised standard of education and training in computer literacy and other areas of technology competence and skill critical to success in employment for youth and young adults ages 14 to 23.
- *Linkages*: Establish and enhance linkages with employers and jobs in the local community for computer literate youth and young adults ages 14 to 23 to improve their chances for competing effectively for employment and to support them in transitioning to the world of work.
- *Statewide Network*: Participate in mutually beneficial collaborative learning and sharing relationships among the leaders of the 11 CCCs.
- *Evaluation*: Throughout the funding period, centers will collect data regarding participation of youth and local residents as well as program outcomes and impacts in order to continually evaluate progress toward their stated objectives. Centers will work with the initiative evaluators and the coordination team to ensure progress toward program and CIOF goals.
- *Community Voice*: Formalize the establishment of a broadly representative CommTAC whose members will build support for the center's work, and assist in advocating for local positions and policies that ensure equitable technology access for low-income communities.
- *Community Technology Resource*: Have linkages to share expertise with and serve as a technology resource for community-based organizations, schools, businesses, and individuals.
- *Policy Development*: Identify, promote, and achieve local, regional, and/or statewide policy reforms that result in improved access to computer technology for people living in low-income communities.
- *Dissemination*: Serve as a role model to others by disseminating lessons learned, success stories, as well as help to educate others on the importance of access to technology and technology education and training for people in low-income community areas.

---

employment linkages with existing programs and increase computer access for local residents. Break-Away Technologies was partnering with the Center for Language Minority Education and Research at the California State University (CSU) Long Beach for curriculum development and support. The CSU Long Beach Foundation was the fiscal agent.

The San Diego Housing Commission teamed up with *Casa Familiar* to increase access to computers, training, and jobs for youth in San Ysidro, a low-income community of 34,000 with no high school or major employers. Building on the San Diego Housing Commission's nationally recognized model for providing computer services to public housing residents,

TABLE 6.2
Organizations Funded to Develop CIOF Community Computing Centers

• *Break-Away Technologies*	• *Karuk Community Development Corporation*
• *Career Resources Development Center*	• *P. F. Bresee Foundation*
	• *Plumas County Department of Health*
• *Central Union High School District*	*Services/Plumas Children's Network (4 sites)*
	• *San Diego Housing Commission/Casa*
• *Continuing Education Division of Santa Barbara City College*	*Familiar*
	• *University of California, Riverside Community Digital Initiative*
• *C. T. Learning, Inc.*	• *Women's Economic Agenda Project*

the commissions and Casa Familiar planned to establish a community computer center in San Ysidro to serve area youth. Casa Familiar, Inc. was originally organized in 1972 under the name of *Trabajadores de la Raza*, San Diego chapter, to serve a specific target population: Spanish-speaking monolingual clients. Over the years, its services and target population expanded to include all of south San Diego's population regardless of racial or ethnic background (however, the demographics of the area are primarily Latino). Through its CIOF community-computing center, Casa Familiar planned to provide access to technology and both English and bilingual training for local residents.

The *University of California, Riverside Community Digital Initiative (CDI)* established a computer laboratory and educational center in the Caesar Chavez Community Center in Riverside. The Center for Virtual Research and the Center for Social and Behavioral Science Research in the College of Humanities, Arts and Social Sciences at the University of California Riverside were directing the initiative in partnership with the Greater Riverside Urban League. The computing center offered structured classes and programs, as well as open lab time. CDI was targeting youth ages 14 to 23 in Riverside's predominantly low-income eastside community, but the laboratory planned to have open hours for all interested users. CDI offered technology training to youth, new mothers and young parents, and people with disabilities through innovative software modules. In cooperation with the local city redevelopment agency, employment enrichment to the people of the eastside community was a priority at CDI.

Plumas County Department of Health Services (PCDHS) is the local public health department for Plumas County, a remote rural area of northern California with a population of 21,000. PCDHS was the fiscal agent and administrator for the Plumas Children's Network (PCN), a local community-based collaboration that planned to implement the Plumas County CIOF project. Because transportation was difficult in the mountainous region, the Plumas CIOF center was spread between four separate sites: Chester, Greenville, Portola, and Quincy—increasing the access to computer technology for all residents. These centers were designed to (a) provide computer education and job training, (b) recruit and engage volunteers to support efforts to improve access to computer technology, training, and jobs, and (c) develop community awareness of the importance of access to computer technology. It was hoped that the CIOF project would help link the four major communities of this remote area and build awareness of the importance of access to technology as a positive resource for educational and economic success for the community.

Career Resource Development Center (CRDC) is a private, nonprofit organization with a 32-year history of providing educational programs to immigrants, refugees, and other disadvantaged populations in San Francisco. Located in San Francisco's Tenderloin district, CRDC partnered with Central City Hospitality House, a homeless and runaway youth services agency, to provide homeless, runaway, immigrant, and refugee youth with computer training, job readiness training, and language instruction. The program emphasized on-site internships, ongoing training, and appropriate job placements. CRDC also worked with other community leaders and organizations to raise awareness of policymakers and the public of how the lack of access to technology affected the health and future of the residents of low-income communities. CRDC primarily served Chinese, African American, Latino, and Southeast Asian youth and young adults. The CIOF lab was designed to support 26 computer workstations and was open four nights a week as well as Saturday afternoons.

C. T. Learning is a Fresno-based, nonprofit organization working to empower residents of low-income communities by developing literacy skills. It was the lead agency in a collaboration of faith-based institutions including Catholic, Episcopal, and Baptist churches, and the Fresno Interfaith Sponsoring Committee. It primarily served African American, Latino, and Southeast Asian youth and young adults. C. T. Learning, Inc. planned to enhance its literacy program by operating a community computer center in central Fresno that increased access to computer technology and built an understanding of the connections between computer literacy and employment for program participants. The project also sought to engage local churches in effective efforts to address the needs of area youth while building church–community leadership and encouraging faith-based institutions to take responsibility for addressing community challenges such as increasing access to technology, training, and jobs for youth and young adults.

The Desert Oasis High School (DOHS) community-computing center, as part of the Central Union High School District, is located in El Centro, a city of 39,000 in the Imperial Valley in the southeast corner of California near the United States–Mexico border. The community-computing center was established at DOHS, an alternative education program for high school students that also served as the site for the District's adult education program. Building on an existing technology-based program, the community computer center targeted at-risk youth ages 14 to 23 to help them improve academic knowledge and develop computer, English, and job skills. Each week the center provided 57 hours of unrestricted access to the 31-station computer center for community members. In addition to the open access services, several training classes and study programs were available, teaching computer skills and job readiness. Through collaborative linkages, students obtained work experience placements, on-the-job training opportunities, or actual employment. A comprehensive computerized English study program, Rosetta Stone, builds confidence for non-English speakers. The center was used by people ages 3 to over 70 to accomplish a wide range of tasks, such as improving academic proficiency, preparing to become a U.S. citizen, completing schoolwork, working on a home or business project, researching on the Internet, or just having fun.

Happy Camp Community Computing Center (HCCCC). The Karuk Tribe of Happy Camp is a federally recognized Indian tribe located in the remote mid-Klamath river area of Siskiyou County, near the Oregon border. The economy of the area had been adversely impacted by the decline of the timber industry, and 85% of its 4,000 residents were considered low-income by federal standards. Through the Karuk Tribe Community Development Corporation, the Karuk tribe and a broad spectrum of community organizations and representatives were working together to develop and implement economic revitalization strategies. Toward this goal, the Karuk Tribe planned to use funding and support from CIOF to provide access to technology and training for local youth and young adults to prepare them for a broad array of occupations. The project offered access to computer technologies to a community previously unserved by any computer facility. During the evaluation, HCCCC staff became enrolled in the California Technology Assistance Project, a statewide educational technology leadership initiative that provides assistance to schools and districts in integrating technology into teaching and learning.

The P. F. Bresee Foundation is a nonprofit organization that is a source of faith, hope, and service to one of the poorest and most densely populated communities in Los Angeles County. Serving the neighborhoods of Koreatown, Pico Union, Westlake, and south central Los Angeles, the organization was working to build a stronger, healthier, safer community for area residents by supporting community and youth development and urban leadership efforts, as well as by increasing access to basic health services. Located in the First Church of the Nazarene, the Bresee Foundation planned to use funding from CIOF to expand the hours of its Cyberhood Computer Education Lab to increase the availability of basic and advanced computer training for young people ages 14 to 23. Operating the lab as a community computer center, the Bresee Foundation would work to increase the access to technology and training, improve school performance, and enhance the job skills and employability of young people. Bresee also (a) provided computer and technology assistance to community organizations serving the residents of central and south Los Angeles, (b) tracked and monitored the availability of technology and training in neighboring schools and city programs, and (c) advocated for increased computer access and training.

The Continuing Education Division of Santa Barbara City College (SBCC) is Santa Barbara's primary public sector provider of computer and technology-related instruction and delivers both English language and bilingual instruction in computer application and vocational education. The division collaborated with the Santa Barbara High School District, the Santa Barbara County Office of Education and Transition House, a local grass roots volunteer-based service agency, to implement the CIOF program in Santa Barbara. Focusing on Santa Barbara's low-income population, the program ran a community-computing center in downtown Santa Barbara at the Schott Continuing Education Center on West Padre Street. SBCC opened a second site at the Wake Continuing Education Center on Turnpike Road. A third center included the Isla Vista Teen Center. These CIOF centers provided critical technology access to people who had no computer at home, or at work. Participants were 70% Latino and all those who came for computer access and/or training were low income.

The Women's Economic Agenda Project (WEAP) is an 18-year-old, community-based nonprofit organization in Oakland led by low-income women of color. It was established to empower poor women to assume leadership and work for economic justice and self-sufficiency. Part of its plan was to use CIOF funding to help lift women out of poverty by helping them develop computer and technology related skills. WEAP offered technology training and access to computers at its Computer and Tele-communications Skills Center. Primarily providing training for young minority women, the center strived to increase access to technology for the residents of the surrounding low-income community, provide computer-related job training to young at-risk women and men, build linkages with employers, increase job placement opportunities, and advocate for equitable technology access.

Developing Program Impact Theory

Once the sites just described were funded by TCWF, staff from each organization participated in discussions about developing a program theory for CIOF. With extensive input from site leaders, TCWF program officers, and the CIOF program coordination team, the evaluation team constructed a guiding program theory of how the CIOF program was presumed to work. Figure 6.1 shows that participation in the CIOF program was believed to lead to improved attitudes toward computers, technology skills, career development knowledge, job search skills, and basic life skills. These acquired skills and knowledge were presumed to facilitate the pursuit of more education, internship opportunities, and better employment options, which in the long-term, were expected to improve participants' health status.

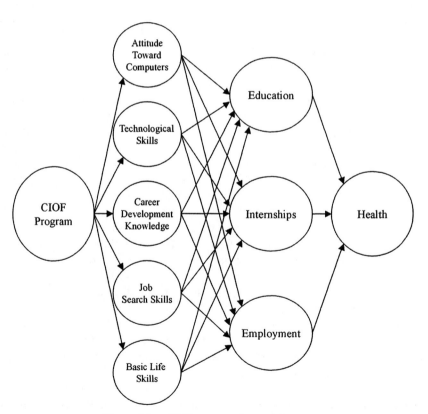

Figure 6.1. CIOF program impact theory.

FORMULATING AND PRIORITIZING EVALUATION QUESTIONS

The program impact theory described was used to develop and prioritize evaluation questions that were agreed on by program stakeholders. More specifically, during the earliest phase of program implementation, leaders from the program coordination team (PCT) and TCWF program officers, as well as leaders from each of the CIOF CCCs and the evaluation team agreed on four core areas that the evaluation would address. Evaluation resources would subsequently be implemented to answer core questions in these four areas:

1. *Program Implementation.* What does it take to set up a vibrant, accessible, relevant, and sustainable community computing center? What does implementation look like? What are the key challenges and success factors to program development and implementation?
2. *Program Service.* Whom are the sites serving (e.g., population characteristics of service recipients)? How many people are served at each site?
3. *Program Impact.* What is the impact of the CIOF program on the 11 sites, their participants, and their communities?

    a. How do technology access and training improve the employment prospects of young people?
    b. What are realistic outcomes in increased computer or technical skills, employment, increased literacy, English acquisition, attainment of GED, or other educational targets?
    c. How does the center affect the participants personally (e.g., self-confidence, motivation, life skills)?
    d. What are the demonstrable payoffs to communities (e.g., increased cohesion, access to technology resources and services, etc.)?

4. *Strengths and Weaknesses of the CIOF Model.* What are the strengths and weaknesses of the specific CIOF models at each of the CCCs (i.e., access, education, resources, and voice)?

These questions were used to guide data collection and reporting efforts.

ANSWERING EVALUATION QUESTIONS

This section illustrates how Step 3 in the program theory-driven evaluation science process played out in the evaluation of CIOF. It is organized around the topics of data collection, key evaluation findings, and consid-erations for future programming. These topics will illustrate once again how the transition from the conceptual Step 1 and Step 2 to the empirical Step 3 unfolded in the program theory-driven evaluation process.

Data Collection

Extensive discussions occurred about how best to collect data to pro-vide a basis for evaluative conclusions for the CIOF evaluation. A com-bination of quantitative and qualitative methods was used to answer the evaluation questions described. For example, extensive standard-ized demographic, center utilization, pretest, posttest, and follow-up data were collected from each site. Various types of qualitative imple-mentation and outcome data were also collected.

Overall, data were collected for over 25,000 program participants, including user data (demographic and background) for 22,729 individuals, center usage data (daily activities and time spent at center) for 12,049 indi-viduals, pretest and posttest data for over 400 individuals, and follow-up interview data for over 200 individuals. Both user and usage data were col-lected for nearly half (47%) of all individuals tracked. Data collected with these measures included demographic and background data, computer-related behaviors inside and outside the CIOF centers, attitudes toward computers, computer skills, reactions to the CIOF centers, and educa-tional and employment outcomes. These data were limited in similar ways to the WNJ data (e.g., one group, pretest–posttest research design) when used to estimate program impact. However, various types of qualitative implementation and outcome data were also collected, including site visit observations and interview data from site leaders and program coordina-tion team members. In addition, more than 80 site progress reports and eight progress reports from the PCT were analyzed for key accomplish-ments and lessons learned.

To support continuous program improvement within the CIOF imple-mentation sites throughout the life of the initiative, the evaluation team prepared and disseminated 113 evaluation reports to CIOF program leaders, PCT members, and TCWF over the 5-year funding period. These included five year-end evaluation reports, four midyear evaluation reports, 92 site reports, four interim reports, and eight miscellaneous eval-uation reports to program grantees. As tools for program monitoring and improvement, these reports documented not only key accomplishments and program activities, but also key program challenges and recommen-dations for addressing challenges. Conference calls and/or face-to-face meetings were held with site leaders, the PCT, and TCWF to discuss each report. In addition, the evaluation team presented key findings and updates on statewide service statistics at each of the biannual CIOF statewide conferences. During these various communications, the evaluation team facilitated discussion on the meaning of the findings and on developing strategies and responses to addressing program recommendations.

Several evaluation tools were also created and disseminated to sites to support evaluation and program capacity building. In addition to their evaluation plans, each site received an evaluation procedures manual that served as a centralized resource for storing evaluation materials such as (a) evaluation planning materials, (b) evaluation communications, (c) draft and finalized measures, (d) evaluation feedback reports, (e) sample technology and portfolio evaluation measures from mainstream efforts, and (f) evaluation training materials prepared by the evaluation team. Overall, pretest and posttest measures were created to measure skill gains for more than 40 different classes across the CIOF network. A key resource initially created by CompuMentor and developed and supported by the evaluation team over time with extensive input from site leaders, was the CIOF Client Tracking System (CCTS). The CCTS is a Microsoft Access-based database that tracks client demographic, background, and center usage data. As a tool for program improvement, the evaluation team designed extensive reporting capabilities into the CCTS to enable program leaders access to program service statistics for convenient and timely reporting and dissemination purposes. The evaluation team created a companion training and user manual to assist sites in understanding and maintaining this client tracking system.

Key Evaluation Findings and Conclusions

1. Program Implementation

- CIOF grantees required more technical assistance and management support than expected. Areas of greatest needs for technical assistance included technology set-up and maintenance, program planning, curriculum development, employment linkages, policy development, and resource procurement.
- Although the complex and ambitious nature of the CIOF model was challenging for sites to address, several leaders reported that they accomplished more than they would have if their programs had a narrower focus.
- Site leaders considered relationship building with community partners as a critical component in order to develop successful programs. Key challenges to establishing and maintaining partnerships

centered on role clarification and during the majority of the first half of their funding period on developing their centers, providing unrestricted access to computers, and developing appropriate levels of staffing and instructional support for center users.

- To be effective, open access required clear operating guidelines and guided activities.
- Attracting, developing, and retaining staff were key program management challenges throughout the funding period. Key lessons centered on the need to (a) identify and attract the right mix of technical and people skills, (b) maintain two to three staff per every 10 users, and (c) utilize and develop volunteer support staff. There was general consensus among program leaders that it was easier to develop technical skills than people skills. Key challenges concerned keeping skills current with changing technology and working with students with diverse levels of needs, skills, and learning capacities.
- Providing access and training services was not always sufficient. Program staff often found themselves helping center users cope with life challenges.
- Securing corporate contributions was easier once sites became operational. When compared to equipment donations (hardware and software), contributions for other resources such as training and maintenance support were much less available.
- The best outreach efforts came from center users. Most center users learned of their center through word-of-mouth referrals.
- Strong technical expertise was essential for creating centers. Most sites had to develop these skills or bring in technical expertise from the outside. See Table 6.3.

2. Program Service

A combined total of 25,036 Californians participated in the 14 CIOF centers over the 4-year period (see Table 6.4; note that one of the 11 grantees developed four CIOF centers). Based on systematic tracking data, sites reached 44.4% ($n = 12,293$) of their combined open-access service goal of 27,705. In addition, they reached 69.8% ($n = 3,418$) of their youth training goal of 4,900 for 14- to 23-year-olds, and 15.1% ($n = 648$) of their youth employment and work experience goal of 4,300 for 14- to 23-year-olds.[1]

---

[1]These figures are based on systematic tracking data that were collected by sites using the CCTS. They do not include additional site estimates of individuals served but not tracked in the database by program staff.

Computer Experience and Usage. Over three fourths of all center users reported having used a computer prior to participating in CIOF centers. Two thirds of the center users reported having access to computers outside of CIOF. Nevertheless, center participants visited their CIOF centers an average of 10 times and spent an average of 1 hour 28 minutes per visit. Overall, center users logged approximately 250,000 visits and more than 370,000 hours of use across the CIOF network of centers.

Table 6.5 summarizes clients' status with prior computer experience. More than three fourths (78.5%, $n = 14,566$) of all respondents ($n = 18,557$) reported having used a computer before visiting a CIOF computing center. Less than one third of respondents, however, said they felt comfortable with computers (25.5%) or had a good deal of expertise with computers (1.3%) at the time of intake.

Center usage data from nine CIOF sites were reported, and revealed usage activity for 12,600 unduplicated users. These participants visited the CIOF centers a total of 154,060 times, or an average of 10.1 visits per person. Projecting these to the total sample, we estimate that users logged more than 250,000 visits. In terms of the time spent in the centers, these users collectively logged 292,871 hours of center use since they began tracking utilization data. The average time per visit was 1 hour 28 minutes. Based on these figures, the projected total number of hours logged by all center users since the program began would be 371,709 hours. The core activities users engaged in were open access (48.4%) and technology classes (26.1%). See Table 6.5.

At time of intake, all respondents were asked whether they had access to computers outside the CIOF center they visited. Among those responding (18,101), 33.5% said they did not have computers available to them outside their CIOF center. Among those who did, just over one third (37.2%) said they had computers available to them in school. Compared to over half (59.4%) of Californians having a computer in the home,[2] just over one fourth (27.6%) of CIOF participants said they had a computer available in their home. In addition, 19.2% said there was a computer available at a friend's or family member's home, compared to 10.9% who had a computer available at another community-based organization. Among other locations, 10.3% had a computer available at work, 1.8% at church, and 6.7% at some "other" location. When asked about the quality of these computers, the data revealed that more than one third of CIOF participants (36.2%) had no support or help to learn how to use the computers at these locations.

---

[2]Source: The California Work and Health Survey (1998), University of California, San Francisco.

TABLE 6.3
Center Implementation Spotlights

Happy Camp Community Computing Center	Desert Oasis High School	Plumas County Health Services
Internet Access	*Building Relevance*	*Adult Appeal*
At the Happy Camp Community Computing Center (HCCCC), access to the Internet was one of the biggest benefits to the community. In their isolated area, there was no local radio, television, or newspaper, and very few books in the public library. Entire high school classes were now able to research topics they would never have had access to before the center was available. Center staff feel this has raised the standard of student work. Similarly, adults were now able to book plane tickets, buy and sell stock, research purchases, send and receive e-mail, compile information, and seek employment through Internet access.	At the Desert Oasis High School center, providing equitable access to technology was much more than a hardware and software issue. Instead, they learned that it was important to provide training and education on all the ways in which technology can be used to improve the quality of life for participants. In support of this, they found that having students conduct self-evaluations of technology use was a valuable tool in raising awareness of technology literacy.	In terms of attracting the target population (youth 14 to 23 years old), center leaders in Plumas underestimated the strong appeal of the program to the adult population and the huge demand that e-mail would create. In response, their centers are working more closely with local high schools and after-school programs to draw the target population into the centers.

Characteristics of Population Served. More than half (52.9%) of CIOF participants were female, more than three fourths (78.4%) were people of color, and just over one third (37%) were youth and young adults in the target age range of 14 to 23 years of age.

Table 6.6 summarizes characteristics of the populations served at each site. Overall, the 14 sites served diverse populations. More than half ($n = 6$) of all CIOF centers served predominantly Latino populations (i.e., greater than 50%). Two sites served more than three fourths who were African American, and one site served predominantly Asian American users. Only two sites served populations in which over half

TABLE 6.4
Number of Clients Served at Each CIOF Center (1997–2001)

| | | | Number of Center Users Tracked by Sites | | |
| | | | | Percent of all users engaged in: | |
CIOF SITES	Total # of Unduplicated Users Served	# and (%) with Usage Tracking Data	Open Access	Class Training	Job Search or Job Training
Break Away Technologies (BAT)	2,540	1,850 (72.8%)	94.1%	2.1%	0%
Casa Familiar (CF)[a]	2,446	0 (0%)	—	—	—
Community Digital Initiative (CDI), UC Riverside	1,178	778 (66.0%)	41.9%	44.3%	5.2%
Community Resource Development Corporation	816	706 (86.5%)	40.0%	54.6%	3.0%
CT Learning (CTL)	2,428	1,996 (82.2%)	52.7%	40.3%	0.6%
Desert Oasis/Central Union HS District (CUHSD)	3,353	3,001 (89.5%)	43.1%	22.6%	26.7%
Happy Camp (HCCCC)[a]	1,234	695 (56.3%)	43.4%	56.6%	0%
Plumas Children's Network (PCN)—4 sites	2,931	2,022 (69.0%)	45.7%	21.4%	17.0%
PF Bresee (PFB)	2,118	291 (13.7%)	33.1%	41.0%	10.2%
Santa Barbara City College (SBCC) [a]	5,311	0 (0%)	—	—	—
Women's Economic Agenda Project (WEAP)	681	598 (87.8%)	48.1%	49.6%	0.1%
TOTAL	25,036	11,937 (46.7%)	49.1%	36.9%	7.0%
Projected total number of unduplicated users (all ages) in each activity			12,293	9,238	1,753
Projected total number of unduplicated users (ages 14–23) in each activity			4,548	3,418	648

[a]These sites did not track center usage data.

were 14- to 23-year-olds, whereas five sites served primarily adult (24 years or older) populations.

All participants were asked to state their primary language and whether or not they spoke English. Among those who responded to this question ($n = 20{,}417$), just over one third (36.4%) indicated that Spanish was their primary language, more than half (58.6%) said English, and 5.0% said "other." According to data submitted by sites to the

TABLE 6.5
Number of Visits and Activities

	Full Sample (n = 12,600)	Ages 14-23 (n =3,480)
Computer Experience at Time of Entry to CIOF:		
Ever Used a Computer before CIOF	78.5%	90.5%
Comfortable with Computers at Intake	25.5%	36.6%
Consolidated Usage (Daily Log) Data:		
% Having Access to Computers Elsewhere	66.5%	74.8%
Number of CIOF Visits Logged	154,060	48,498
Projected Total Number of CIOF Visits*	250,000	109,407
Average # of Visits Per Person**	10.1	13.9
Daily Usage ... Percent engaged in =		
Class	26.1%	30.9%
Open Access	48.4%	53.8%
Job Readiness Training	6.7%	9.4%
Job Search Assistance	1.6%	1.8%
Work Experience	.3%	.5%
Youth Program	1.5%	1.9%
School/Work-based Activity	3.2%	6.2%
Special Program	1.0%	1.3%
Basic Skills	1.7%	1.2%
Other	3.5%	2.8%
Hours of Center Use	292,871	70,094
Average Time Per Visit	1 hour 28 minutes	1 hour 6 minutes
Projected Total Hours of Use for All Users***	371,709	120,378
Percent engaged in 1 activity per visit	69.2%	60.1%
Percent engaged in 2 or > activities per visit	30.8%	39.9%

*Totals are projected from the sample providing data (n = 12,600) to the overall sample (n = 25,036) based on average number of visits per person.

**Average number of visits reflects client tracking data from nine sites that submitted daily usage data, and one site (SBCC) that only reported the average number of visits for its center.

***Totals are projected from the sample providing data (n = 12,600) to the overall sample (n = 25,036) based on the average amount of time logged per person.

evaluation team, several CIOF sites served primarily Spanish-speaking populations: Santa Barbara City College (63.8%), Casa Familiar (51.6%), CT Learning (51.2%), and Desert Oasis (45.2%). In addition, more than one fourth of those at P.F. Bresee (28.7%) were primarily Spanish speakers. Fewer than 10% of all users (7.9%) did not speak English.

3. Program Impact

Building computer technology literacy was a central goal of CIOF programs. Developing curriculum and technology training programs, however, proved to be much more challenging for program leaders than
anticipated. Contrary to expectation, it took most sites up to 3 years to
develop program-learning objectives. In part, this was due to centers'
initial focus on developing centers and open-access periods. However,
there were also numerous obstacles to overcome in order to develop
programming in this area. The key challenges to developing technology
curricula included:

- Lack of experience, training, and guidance in creating a program
  curriculum.
- Process of writing down learning objectives—what they wanted
  students to know and be able to do.
- Gathering and developing culturally relevant instructional
  resources to be used as teaching aids.
- Turnover among staff and instructors.
- Locating and hiring instructors having the right mix of technology
  and people skills.
- Competing with the private sector for employees.
- Available time and resources did not match program develop-
  ment needs in this area.
- Training was staff-intensive (two to three qualified instructors for
  every 10 students).
- Attracting and "hooking" youth and young adults into formal
  training experiences.
- Unreliable partnerships with other service providers—did not
  result in anticipated client flow or additional instructor resources.
- Meeting training needs for individuals coming in with different
  learning needs, goals, and capacities.
- Lack of software and training manuals in Spanish or other languages.
- Developing internal capacity with respect to technology knowledge and skills.
- Keeping ahead of the pace of learning and keeping current with
  the latest technology.

Although sites offered learning experiences in many areas, collectively,
they developed learning objectives for 17 different core-technology content areas as part of their program evaluation. These represent the core
skills that sites have identified as most important in fostering employment and educational opportunities for target constituents, youth 14 to
23 years of age. The content areas include: Word Processing (selected by

TABLE 6.6
Demographics of Populations

SITE**	Total Served	Gender (n = 21,006)		Race & Ethnicity (n = 21,024)						Age Group (n = 21,276)		
		Female	Male	African Amer.	Asian Amer.	European Amer.	Latino/a	Native Amer.	Other	<14	14–23	24+
BAT	2,540	55.5%	44.5%	87.1%	0.6%	2.4%	8.3%	0.4%	1.3%	7.2%	29.4%	63.4%
CASA	2,446	48.1%	51.9%	1.6%	0.3%	2.0%	94.1%	0.8%	1.4%	33.5%	46.6%	19.9%
CDI	1,178	54.9%	45.1%	26.2%	2.6%	10.1%	58.9%	0.3%	1.9%	9.8%	49.6%	40.6%
CRDC	816	56.2%	43.8%	7.9%	63.8%	16.5%	5.0%	2.3%	4.5%	9.0%	26.5%	64.5%
CTL	2,428	49.5%	50.5%	5.6%	1.3%	3.5%	86.4%	0.4%	2.9%	32.0%	35.0%	33.0%
DOHS	3,353	62.5%	37.5%	2.3%	0.7%	8.5%	85.1%	0.3%	3.0%	13.6%	43.7%	42.7%
HCCCC	1,234	52.7%	47.3%	0.1%	0.2%	58.7%	3.5%	33.8%	3.8%	14.9%	27.9%	57.1%
PCN	2,931	59.1%	40.9%	1.0%	0.7%	85.9%	4.7%	6.4%	1.3%	11.0%	23.3%	65.7%
PFB	2,118	38.6%	61.4%	29.2%	4.8%	3.0%	56.6%	0.1%	6.3%	14.4%	71.2%	14.4%
SBCC	5,311	48.6%	51.4%	1.7%	2.1%	19.5%	73.3%	0.5%	2.9%	13.8%	25.3%	60.9%
WEAP	681	64.8%	35.2%	73.5%	6.3%	5.8%	12.5%	0.9%	1.0%	2.2%	53.4%	44.4%
TOTAL	25,036	52.9%	47.1%	11.8%	3.0%	21.6%	57.7%	3.1%	2.7%	16.3%	37.0%	46.7%

*Figures in **bold** highlight key populations served.**BAT = Break-Away Technologies; PFB = P.F. Bresee; CASA = Casa Familiar; CRDC = Career Resource Development Center; CTL = C.T. Learning; DOHS = Desert Oasis High School; HCCCC = Happy Camp; PCN = Plumas Children's Network (overall, including Chester, Greenville, Portola, and Quincy); CDI = UC Riverside; SBCC = Santa Barbara City College; WEAP = Women's Economic Agenda Project.

five sites), Spreadsheets (five sites), Computer Graphics (four sites), Keyboarding (four sites), Introduction to Computers and Windows (three sites), Internet (three sites), and Internet and e-mail (two sites). The remaining content areas were selected as core curricular areas by only one site each: Office Automation, Publisher, Computer Basics, Advanced Multimedia, Business I and Business II, Web Design and Net Basics, Computer Repair and Maintenance, CalWorks English, Job Readiness, and PowerPoint. Overall, the number of areas selected ranged from one to six per site. Based on the learning objectives submitted by sites, the evaluation team developed pretest and posttest measures to assess skill gains for over 40 different classes covering these topics.

Some key lessons pertaining to classes included (a) the importance of scheduling classes at times convenient to the target population, (b) offering short and frequent lessons, (c) providing project-based lessons, and (d) using visual textbooks or handout materials that were not intimidating. Some sites even found that what they called their classes (e.g., workshops vs. class; open class vs. open access) was important in attracting the target populations. Finally, a number of site leaders noted that relationship building within classes was central to creating a positive learning environment. More specifically, they noted that finding out what really matters to center users and having students work together were key to engaging them in using computers and helping them develop computer and job-related skills. Making a personal connection was also attributed to increasing the likelihood that center users would return for future classes. One center reported benefiting from making follow-up calls to determine why students were not attending class. They felt this individualized attention let them know the personnel in the center were interested in their progress.

Knowledge and Skill Gains.   The data collected revealed increases in the frequency of computer use, diminished fears of computers and feelings of being marginalized with respect to technology, and gains in the proportion of students who mastered targeted skills over time. However, these findings must be viewed in light of the small sample size. See Table 6.7.

Technical Skill Gains at Follow-Up.   Individuals with whom the evaluation team conducted follow-up interviews noted that they gained multiple types of new technology skills. Two thirds (66.7%) mentioned that they learned how to use specific types of software programs such as Word, Excel, or Publisher. One third (33.8%) reported gaining Internet and e-mail skills, and more than one third (36.3%) reported gaining a variety of general skills such as typing and computer basics (e.g., how computers work, how to operate computers). Two percent mentioned basic skills, such as math skills. Only 3.5% said they did not build any new skills at the centers. See Table 6.8.

Employment Status.   When asked about their employment status at intake, 34.3% of all respondents ($n = 14,292$) said they were currently employed at time of intake. More than half (59.9%) said they were unemployed, 1.6% said they were underemployed, and 4.3% were retired. Among those who were employed at time of intake, 15.4% held two or more jobs, 59.4% worked full time, 40.6% worked part time, and 40.4% were in school at the time. In terms of their jobs, the average hours worked per week was 32.7, and the average hourly wage was $9.73 per hour.

Just over two-fifths (40.8%) of all respondents ($n = 7,572$) said they were seeking employment at the time of intake, including 38.3% of 14- to 23-year-olds. Over one fourth (29.3%) of those who were employed at intake said they were currently seeking employment. See Table 6.9.

Employment Outcomes.   Most programs centered on skill building and employment preparation in order to develop links leading to employment. Centers viewed successful educational attainment as a priority for ensuring improved employment opportunities for the target population. Developing employment links was very challenging for centers. Most program stakeholders lacked appropriate experience, knowledge, and/or resources to create solid programming around job development, placement, and tracking employment outcomes. It should be noted that due to the challenging nature of this goal and the sites' focus on technology access and training, this goal was de-emphasized by TCWF over time.

Several types of employment activities were developed at several of the centers, including job search assistance, career exploration, job readiness, resume writing, interviewing skills, and job skills. Daily client tracking data showed that 7% of all users were engaged in employment-related activities at the centers. In comparison, 37% were involved in technology skill building classes.

Program leaders were asked to identify individuals they considered to be "success stories" to participate in follow-up interviews with the evaluation team. Overall, 202 respondents were asked about what they did in their center and how they benefited from their participation in CIOF programs. Respondents were asked both general questions about how they benefited as well as specific employment-related experiences. Employment results are summarized in Table 6.10.

More than one fourth of respondents reported participating in job-readiness classes (28%) and job-search activities (31%) at their CIOF

TABLE 6.7
Technical Skill Gains and Attitudes Toward Technology

	Pretest	Posttest
Computer Use		
Frequency of Computer Use Outside CIOF	(n = 450)	(n = 264)
4+ times/week	27.8% (126)	36.0% (95)
2–3 times/week	20.7% (93)	28.4% (75)
Once/week	18.4% (83)	16.7% (44)
1–3 times/month	10.7% (48)	9.1% (24)
Less frequently	10.7% (48)	10.2% (27)
Never used	11.8% (53)	4.2% (11)
Attitudes Toward Computers		
Fear of Computers	(n = 335)	(n = 132)
Low	54.9% (184)	74.2% (98)
Moderate	41.8% (140)	25.0% (33)
High	3.3% (11)	.8% (1)
Computer Marginalization	(n = 346)	(n = 136)
Low	72.2% (275)	82.4% (112)
Moderate	21.6% (75)	16.1% (22)
High	1.2% (4)	1.5% (2)
Technology Skill Gains		
Mastery of Target Technology Skills		
Percent able to do skills on own	32.3%	63.0%

centers. Over half of all respondents said they developed job-related technology skills (61%), researched skills needed for jobs (60%), discovered career interests (57%), learned how to conduct a job search (56%), and learned how to develop a resume (51%).

When asked more generally how they personally benefited from participating in CIOF, responses ranged from building new computer knowledge and skills to changing their attitudes toward computers. Key employment-related benefits are now summarized.

- Found a job (6%).
- Applied new skills at work (6%).
- Increased job search self-efficacy (4.5%).
- Used for job search (3.5%).
- Improved career opportunities (3.5%).
- Developed resumes and interviews (2.5%).

TABLE 6.8
Follow-Up Self-Reported Technical Skill Gains

	All Respondents (n = 201)	14–23 Year Olds (n = 89)
General Programs	66.7% (134)	89.9% (80)
Excel	15.9% (32)	15.7% (14)
Reports/calendars	3.0% (6)	3.4% (3)
PowerPoint	8.9% (18)	9.0% (8)
Letters	4.0% (8)	3.4% (3)
PhotoShop	9.9% (20)	11.2% (10)
Word	17.9% (36)	15.7% (14)
Access	4.0% (8)	5.6% (5)
Publisher	2.0% (4)	2.2% (2)
General Programs	17.4% (35)	21.3 (19)
Graphic Design	1.5% (3)	2.2% (2)
Internet & E-mail	33.8% (68)	39.3% (35)
Internet	20.4% (41)	22.5% (20)
E-mail	6.5% (13)	3.4% (3)
Web design	7.0% (14)	11.2% (10)
HTML	1.5% (3)	2.2% (2)
General	36.3% (73)	48.3% (43)
Computer basics	10.4% (21)	12.4% (11)
Typing	22.4% (45)	29.2% (26)
Communication skills	2.0% (4)	4.5% (4)
School-related skills	1.5% (3)	2.2% (2)
Basic Skills	2.0% (4)	1.1% (1)
Installation & Use of Software	1.0% (2)	2.2% (2)
Other	1.0% (2)	0% (0)
None	3.5% (7)	4.5% (4)

Program leaders also attempted to track youth employment and work experiences and reported on these in semiannual progress reports. Sites collectively reported that more than 2,000, 14- to 23-year-olds were placed in work experiences during the program period. Work experiences included employment and job readiness training activities, internships, volunteer activities, and jobs. Unfortunately, sites did not

TABLE 6.9
Employment Status

Employment Status	Full Sample	All Adults (24+)	14–23 Year Olds
Percent employed	34.3%	46.4%	24.4%
Unemployed	59.9%	45.6%	74.0%
Underemployed	1.6%	1.2%	1.7%
Retired	4.3%	6.9%	0%
Percent with 2+ jobs	15.4%	14.1%	15.9%
Percent seeking employment	40.8%	41.2%	38.3%

conduct systematic follow-up or tracking of employment experiences of their participants following participation in CIOF. Therefore, it remains unclear as to how many participants actually obtained jobs or internships. A key weakness in this area was the lack of clear criteria or standards for achieving employment goals. This was due, in part, to evolving TCWF priorities that focused more on training than on employment. An equally important challenge was the lack of systematic employment follow-up tracking to assess employment gains.

Educational Status. Nearly two thirds (61.6%) of all participants and more than four fifths (85%) of those between 14 and 23 years of age were students at the time they began using CIOF centers. Overall, more than four fifths held a high school degree (or equivalent) or had fewer years of formal education.

As expected, the majority of all participants (61.1%) were students, including more than four fifths (85.0%) of those ages 14 to 23. Among respondents who provided education data ($n = 18,518$), 84.3% held a high school degree (or equivalent) or less. More specifically, 1.7% had no schooling, 22.3% had less than eight years of schooling, 32.6% had 8 to 11 years, and 27.7% held a high school degree or equivalent. In terms of advanced training, very few (1.1%) had technical or certificate program training. In terms of higher education, 5.4% held 2-year degrees, 7.0% held 4-year degrees, and 2.2% held a postgraduate degree.

Educational Benefits. More than four fifths (82%) felt their educational opportunities had improved as a result of CIOF. The key ways in which they benefited centered on being able to do homework, school projects and research at the centers (25.4%), improving performance at school (19.9%), enhancing their motivation to go on or do better in school (16.9%), and developing basic skills that help in school, such as

TABLE 6.10
Follow-up Employment Outcomes

Employment Activities	Full Sample	Adults (24+)	14–23 Year Olds
Classes or training programs engaged in:	(n = 161)	(n = 62)	(n = 85)
Job readiness	28.0% (45)	24.2% (15)	35.3% (30)
Job search assistance	10.6% (7)	6.5% (4)	12.9% (11)
Work experience	5.0% (8)	8.1% (5)	5.9% (5)
Job-related projects worked on at center:	(n = 157)	(n = 49)	(n = 72)
Job search activities	30.6% (48)	14.3% (7)	8.3% (6)
Resumes & cover letters	21.7% (34)	21.7% (10)	24.7% (18)
Work-related experiences at CIOF center:	(n = 162)	(n = 64)	(n = 89)
Developed job-related technology skills	61.1% (99)	39.1% (25)	64.0% (57)
Researched needed job skills	59.9% (97)	45.3% (29)	56.2% (50)
Learned career interests	57.4% (93)	39.1% (25)	52.8% (47)
Learned how to conduct a job search	55.6% (90)	35.9% (23)	56.2% (50)
Developed a resume	51.2% (83)	40.6% (26)	51.7% (46)
Developed interview skills	34.6% (56)	20.3% (13)	37.1% (33)
Got internship or work experience	30.2% (49)	10.9% (7)	39.3% (35)
Got a job	27.8% (45)	18.8% (12)	30.3% (27)
Personal benefits that were employment related:	(n = 200)	(n = 64)	(n = 89)
Found a job	6.0% (12)	4.7% (3)	9.0% (8)
Applied new skills at work/ at my job	6.0% (12)	6.3% (4)	6.7% (6)
Increased my job search self-efficacy	4.5% (9)	6.3% (4)	4.5% (4)
Used for job search, finding job	3.5% (7)	7.8% (5)	2.2% (2)
Improved career opportunities	3.5% (7)	3.1% (2)	2.2% (2)
Improved resumes and interview skills	2.5% (5)	1.6% (1)	3.4% (3)
Work-related use in daily life:	(n = 187)	(n = 59)	(n = 83)
Facilitated job performance	18.2% (34)	25.4% (15)	18.1% (15)
Improved job skills	13.9% (26)	23.7% (14)	6.0% (5)
Performed job search	16.6% (31)	20.3% (12)	18.1% (15)
Computer use outside center:	(n = 192)	(n = 64)	(n = 89)
Used a computer at work	18.2% (35)	23.4% (15)	19.1% (17)

typing. A few respondents (2.5%) said they were able to take or pass the GED as a result of using their CIOF center.

Program leaders were asked to identify individuals they considered to be success stories to participate in follow-up interviews with the evaluation team. Overall, 202 respondents were asked about what they did in their center and how they benefited from their participation in CIOF programs. Respondents were asked both general questions about how they benefited as well as questions pertaining to educational outcomes. Findings pertaining to educational outcomes are summarized in Table 6.11.

Although less than 10% of all respondents were engaged in structured school- or work-based activities at their centers, one fifth used their centers to work on homework and school projects. When asked how they used computers in their daily life, nearly one third (30.5%) of all respondents, and more than two fifths (43%) of 14- to 23-year-olds said they used computers to help improve their school performance.

Personal Benefits.   The key ways in which respondents benefited personally included changing their attitudes toward computers (31.5%), gaining new computer knowledge and skills (28.5%), gaining new employment and job-related benefits (26%), being able to use and apply computer skills (18%), and changing their outlook and/or aspirations (13.5%). Other benefits included Internet and e-mail, personal life benefits, helping others, enjoying the center, educational benefits, and improving basic skills.

Respondents were asked open-ended questions about what they liked most and how they benefited from participating in CIOF centers. All comments were content coded and grouped by similar responses. See Table 6.12.

Attitudes Toward Computers.   When asked how they benefited personally from using their CIOF centers, nearly one third (31.5%) of all respondents said it changed their attitudes toward computers. For many, this meant they experienced an increase in their confidence in using computers. As one respondent noted, "I feel more confident about myself because I am able to do things I wasn't able to do before." Other respondents noted they had more positive attitudes toward computers. For example, some who had not liked computers previously now liked them. Other ways that attitudes changed included a decrease in the fear of using computers and an increase in confidence in computer use. As one user stated, "I now feel more confident using the computer. Before, I was afraid of damaging computers."

New Computer Knowledge and Skills.   The second key way in which respondents benefited personally involved learning new knowledge and skills in technology. Overall, most commented that they gained new knowledge or increased their technology knowledge. The statement, "It gave me a lot of knowledge that I didn't have before," was reflected in

TABLE 6.11
Follow-Up Educational Benefits

	All Respondents	14-23 Year Olds
Education status	(n = 18,518)	(n = 6,650)
Less than high school	56.6%	78.6%
High school/GED equivalent	27.7%	18.4%
Some college or higher	15.7%	2.7%
Currently a student	61.1%	85.0%
Educational uses		
Classes or training programs engaged in:	(n = 161)	(n = 72)
School/work-based activity	3.7% (6)	5.6% (4)
Basic skills	7.4% (12)	4.2% (3)
Educational projects worked on in center:	(n = 157)	(n = 70)
Homework and school projects	20.4% (32)	22.9% (16)
Educational use in daily life:	(n = 187)	(n = 79)
Improve school performance	30.5% (57)	43.0% (34)
Computer use outside of CIOF	(n = 192)	(n = 79)
Home	48.9 % (94)	46.8% (37)
School	27.6% (53)	36.7% (29)
Library	7.8% (15)	3.8 % (3)
Educational benefits		
Improve/enhance educational opportunities	(n = 149)	(n = 71)
% Yes	81.9% (122)	70.4% (50)
Ways improved educational opportunities	(n = 201)	(n = 89)
Improved performance	19.9% (40)	24.7% (22)
Enhanced motivation	16.9% (34)	13.5% (12)
Developed basic skills	16.9% (34)	18.0% (16)
Application (homework, project, research)	25.4% (51)	34.8% (31)
Took or passed GED	2.5% (5)	3.4% (3)
Other	16.9% (34)	14.6% (13)

many of the comments. Others claimed that they learned more about computers that enabled them to "do more things now." Several respondents said they got new or better skills. Furthermore, several noted that they now had an increased understanding of computers. As one respondent noted, "It increased my knowledge of computers, not only how it works, but why it works."

Employment-Related Benefits.   The third key way in which respondents benefited from using the centers consisted of employment benefits. Several respondents ($n = 12$) noted they found a job as a result of using a center. Others noted that they were able to apply their new skills to their jobs in ways that improved their jobs. As one respondent noted, "I learned a lot of programs that now I need for the job." Some respondents noted they benefited by learning to use their center to search for jobs; others stated it helped them increase their job search self-efficacy, that is, it made them feel more confident to get a job. Other employment-related benefits noted by respondents were that they now had better career opportunities, knew how to develop a resume and be effective in job interviews.

Use and Application of Technology.   The fourth way respondents benefited was in being able to use computers to do new things. For some, this meant they were now able to use computers or use them more often. Many stated, however, that they developed better and faster typing skills, which enabled them to use computers more efficiently. One respondent said, "I learned a lot about how to use the programs, and learned more skills with navigating around the computer."

Changed Outlook and/or Aspirations.   The fifth key way respondents benefited was to improve their motivation or life aspirations. Many said that the centers opened up new doors to them. Sample comments include: "It helped open a window to computers," and "I used to not know what I wanted to do with my life. I'd like to do this for the rest of my life." Other respondents noted that using the center changed their lives, so they now spend more time on a computer than watching television. Other comments indicated that using the center increased their motivation to learn more about computers and to keep up with changing technology. Two respondents even commented that it changed how they viewed themselves: "Now I am a computer person," said one. Other key benefits included being able to use the Internet and e-mail, helping and teaching others about computers, being able to communicate with family and friends, having access to computers, enjoying the center, and being able to do homework at the center.

TABLE 6.12
How Respondents Personally Benefited from CIOF

	All Respondents	14-23 Year Olds
How Respondents Personally Benefited from CIOF		
	(n = 200)	*(n = 89)*
Attitudes toward computers	31.5% (63)	21.3% (19)
Computer knowledge and skills	28.5% (57)	34.5% (30)
Employment and related benefits	26.0% (52)	28.1% (25)
Use/application	18.0% (36)	13.5% (12)
Changed their outlook and/ or aspirations	13.5% (27)	9.0% (8)
Internet and e-mail	10.0% (20)	10.1% (9)
Personal life benefits	9.0% (18)	3.4% (3)
Helping others	9.0% (18)	7.9% (7)
Liked center and computers	7.5% (15)	1.1% (1)
Educational benefits	6.5% (13)	7.9% (7)
Improved basic skills	3.0% (6)	2.2% (2)
General feelings of benefiting a lot	2.5% (5)	1.1% (1)
Other	3.5% (7)	3.4% (3)
N/A, don't know	*1.5% (3)*	*2.2% (2)*
What They Liked Most		
	(n = 190)	*(n = 89)*
Center environment	26.8% (51)	12.4% (11)
Staff and instructors	25.8% (49)	16.8% (15)
Access to computers and center resources	23.7% (45)	25.8% (23)
Structure of center	12.6% (24)	12.4% (11)
Educational benefits	11.6% (22)	12.4% (11)
Classes and training	7.4% (14)	6.7% (6)
Projects and activities	5.8% (11)	7.9% (7)
Not applicable	12.1% (23)	11.2% (10)
Other	8.9% (17)	12.4% (11)
None	2.6% (5)	4.5% (4)

What Center Users Liked Most. When asked what they liked most about their CIOF centers, more than a quarter of respondents identified their center environment (26.8%) and their staff and instructors (25.8%). More than one fifth (23.7%) cited that having access to computers and resources at their center was what they liked most. Other factors respondents liked included the location and schedule of their centers (12.6%), educational benefits from using their center (11.6%), classes and training (7.4%), and working on projects and activities at their center (5.8%).

Center Environment. The most frequently cited factors that respondents liked had to do with the center environment. Factors such as a warm, friendly environment and staff were mentioned most. Sample comments include, "I like that everyone that works there is friendly," and "I enjoy the atmosphere, that everyone is welcome, and that instructors give individual attention." Other aspects of the environment that respondents liked included having a self-paced environment, a good working environment, and a place to come where they could fulfill social needs. As one respondent noted, "I like that people can learn at their own pace." Another commented, "You can do anything, it's quiet; it's a good place to work."

Staff and Instructors. The second factor that respondents most liked concerned the people who ran the centers and helped them learn about computers. Attributes most frequently mentioned about instructors and staff included that they were helpful, friendly, and that they gave personal help. Many also commented that their instructors were great, patient, and flexible. A number of respondents also said that they liked the fact that there was always help available to them when they visited the center.

Access to Computers. The third area mentioned as being especially favorable had to do with access to computers and technology resources provided at the centers. Respondents said that they liked having access to computers, or having free access to computers. In addition, they mentioned having access to the Internet and Web sites, as well as having access to software programs that they could not get elsewhere.

Structure of the Centers. Several respondents noted that they "liked everything" about their centers. More specific comments centered on convenient class schedules and hours of use, good locations, and clear-cut rules.

Educational Benefits. The key types of educational benefits that respondents liked consisted of learning new things and being able to do research and homework at the center.

CIOF as a Community Resource. Above and beyond having access to hardware and software (21.7%), access to staff and instructors (33%) was the key resource respondents felt they received from their CIOF centers. Other factors that respondents said their centers provided that they could not get elsewhere included: Internet access (18.5%), access to more or better programs (18.5%), and a caring center environment

(18.5%). Additional resources were access to more peripherals (7.6%), convenient schedules and location (7.6%), and classes and training (7.6%). See Table 6.13.

Staff and Instructors.   One third of respondents identified their staff and instructors as the key resource provided by their center. In addition to having instructors, having available help, tutoring and advice, and individualized help were key factors mentioned.

Access to Center and Computers.   One fifth of respondents said that having access to a computer center and computers was a key resource they gained from CIOF. Many noted that it was having free or low-cost access that was important, whereas others noted that they had access to more computers.

Internet Access.   Nearly one fifth of all respondents said free access to the Internet, better access to the Internet, or simply access in itself was a key resource offered by their center.

Access to More and Better Programs.   Respondents differentiated between access to computers in general, and access to more and better programs, in particular. Here, respondents cited that centers offered better resources than other locations by offering a broader selection of programs or the latest versions of programs. Others mentioned having access to specific types of programs, such as PhotoShop, PowerPoint, Publisher, and Word.

Warm Center Environment.   Another factor viewed as a key resource concerned the centers' warm, friendly environments. Comments reflected that respondents felt comfortable at their center, that their staff was friendly, and that the center had a good social environment into which they could plug. Others noted that their centers were places that offered pressure-free and relaxed environments.

Other Key Resources Developed.   In addition to providing access to computers and training in technology, other key resources developed by CIOF centers included the CIOF policy agenda, CIOF newsletters, CIOF Web site, and CIOF toolkits. Each of these inform those interested in community technology about lessons learned, key activities of the centers, and policy implications for community technology centers. Furthermore, these resources are made available to the public at no cost.

TABLE 6.13
Respondents' View of CIOF Program's Value Added

What Respondents Get From CIOF but not Elsewhere	All Respondents (n = 157)	14-23 Year Olds (n = 89)
Staff and instructors	33.1% (52)	32.6% (29)
Access to center and computers	21.7% (34)	14.6% (13)
Internet access	18.5% (29)	12.4% (11)
Access to more, better programs	18.5% (29)	15.7% (14)
Warm center environment	18.5% (29)	9.0% (8)
Access to more peripherals	7.6% (12)	7.9% (7)
Convenient schedule and location	7.6% (12)	5.6% (5)
Classes, training	7.6% (12)	4.5% (4)
Other	4.5% (7)	1.1% (1)
Don't know	7.6% (12)	5.6% (5)
Nothing	3.8% (6)	5.6% (5)

4. Strengths and Weaknesses of the CIOF Model

Technology Education. Because most centers' curricula were not institutionalized in writing, it was not clear which curriculum methods or resources were most helpful in assisting mastery of core learning objectives.

It seems critical for centers to be able to develop a curriculum that supports clear, measurable objectives. Clarifying what it is that students should know and be able to do is the first step in developing and institutionalizing program curricula. It also provides a beacon for measuring learning outcomes and demonstrating technology skill gains. Unfortunately, staff turnover usually resulted in a loss of curriculum and learning objectives when they were not institutionalized in writing.

A goal of the CIOF model was to raise the standard of education and training in computer literacy and other areas of technology competence critical to employment of youth and young adults. A key weakness of CIOF efforts was a failure to integrate more mainstream technology literacy standards into the curriculum. Overall, CIOF targeted development of basic computer technology skills, and in some cases, more in-depth skill building. In general, many of the CIOF centers offered classes covering each topic in one to two class sessions lasting approximately 1 to 2 hours. Some centers offered more extensive training over longer periods of time. Although developing foundational technology skills serves to fill a key gap in low-income communities, developing job-ready proficiency in technology skills requires greater depths of training and practice. Key factors important for designing effective technology training in low-income communities include (a) addressing low literacy rates, (b) developing culturally relevant curricula, and (c) overcoming people's fear of computers. Key challenges include finding staff with the right mix of technical and people skills and meeting ongoing staff development needs (e.g., keeping up with the pace of technological change).

Employment Linkages.   Many program stakeholders questioned the appropriateness of employment as a key goal for youth 14 to 23 years of age. In addition, many lacked job development backgrounds and experience that would enable them to identify the best ways to prepare for and link participants to jobs.

Overall, sites did a better job at providing employment-relevant training than job search and placement assistance. Many program leaders viewed all training as job-relevant, and did not recognize or address the need to match technology skills needed by employers with training and curricular activities. Related to this, they did not make sufficient effort to identify the employment needs of individuals served and local employers. A key lesson learned was that technology is not an end in itself. Instead, technology should be used as a tool that helps to achieve individual and organizational employment goals. A key challenge was balancing employment with technology training goals. Many centers struggled to develop and implement employment programs into existing programs at the same time they were creating technology access and training programs. More successful sites in this area hired dedicated individuals to build and manage their job development program components.

Policy Development.   One of the most notable achievements in the CIOF program was the centers' involvement in the policy-making process around technology access.

Prior to CIOF, most site leaders had little or no experience in policy advocacy. Although it was difficult to articulate and develop policy programming at the individual program level, site leaders realized early on that policy decisions at local, regional, and state levels would be critical to their ability to strengthen and sustain their centers. To achieve this required learning the basics of policy influence and development. This was accomplished by working closely with PCT members who were selected, in part, because of their expertise in policy. In addition to having ready access to PCT policy skills, CIOF centers were connected to other policy groups by the PCT and involved at all levels in discussions on policy influence and advocacy. As a result, not only did CIOF centers develop a policy agenda to take to policy makers, but they also gained access to policymakers on several occasions and presented their agenda to them and to others interested in community technology issues.

Resource Procurement. The CIOF coordination team was instrumental in securing more than $1.6 million in corporate sponsorship for CIOF. Overall, CIOF was successful in establishing 15 corporate partners with resource commitments totaling more than $1.6 million. CIOF partners included: *Premier Partners* ($100,000+): Microsoft Corporation, Adobe Systems, Inc., AT&T, Pacific Bell, Computers for Kids/Futurenet Online Inc., Mattel Foundation; *Connected Partners* ($10,000+): bd Systems, Inc., J.M. Long Foundation, Sony Pictures Entertainment (BreakAway's 200 by 2000 Initiative), The Learning Company; *Partners* ($5,000+): O'Melveny & Myers, Software.com, Inc., Southern California Edison; and *Contributors* (less than $5,000): Legacy Software, Alan Friedenthal.

CONSIDERATIONS FOR FUTURE PROGRAMMING

This section highlights key information and/or lessons learned from the evaluation that might improve future programming in this area. Key considerations included providing professional program planning and development assistance, facilitating strong program leadership, and providing core curriculum modules.

Professional Program Planning and Development Assistance. Most center leaders did not recognize the value of developing and having a technology-training plan until their programs were in place. In many cases, they were not able to identify existing and anticipated community needs and resources. For example, key challenges were (a) identifying how technology training fit into their larger organizational goals, (b) how to integrate new program efforts with existing activities, and (c) how to achieve program goals that required new expertise (e.g., job placement). One implication of these findings is that substantial support for professional assistance in program planning and development may be necessary for centers to overcome key challenges in a timely manner. Minimizing geographical distance between technical assistance providers and grantees is also important. Feedback from grantees suggests that providing mechanisms to identify and link grantees to local service providers may be beneficial. In addition, different models that allow for greater flexibility for provisions of professional program planning and developmental assistance may be necessary, depending on specific needs of grantees.

Facilitate Strong Program Leadership.   Program leaders often struggled with what they felt were unclear program expectations from TCWF. This was due, in part, to how TCWF structured the grant. More specifically, feedback from grantees revealed that they were uncomfortable with the lack of predetermined standards for what they were funded to accomplish. There was a general discomfort with relying on their own expertise to make decisions in developing their innovative programs. Although there was general agreement that center leaders and staff usually knew best whom they wanted to serve, what they wanted to achieve, and how the use of technology would make that happen, they still desired and needed advice and direction. In the case of CIOF, there was a learning curve that sites needed to go through to understand and accept this role responsibility. Program coordination team members, in turn, also learned the importance of the need to provide support and appropriate levels of direction and guidance. Opportunities to meet and talk with others dealing with these concerns were reported to be helpful by grantees.

Provide Core Curriculum Modules.   Many CIOF sites lacked requisite skills for developing technology curricula. Although grantees requested additional support in curriculum development early on, the PCT did not have strong skills in this area. A key lesson was the importance of providing sites with generic technology curricula and learning objectives that can be adapted for local use. Although there were many technology-training resources on the market, grantees felt that many lacked the appropriate levels of language, literacy, and cultural relevance needed in low-income communities. Making generic resources available would show program instructors where to begin so they could avoid reinventing the wheel with respect to technology curriculum. Selecting technical assistance providers with skills in this area could greatly enhance and expedite the curriculum development process.

CONCLUSION

The application of program theory-driven evaluation science to developing and improving a program that created 14 community computing centers in low-income communities, and 11 organizations funded to deliver the program to over 30,000 youth and young adults in the state of California, has been presented in this chapter. A program impact theory was developed collaboratively with stakeholders, and used to focus evaluation resources on answering questions in four main areas: program implementation, program service, program impact, and strengths and weaknesses of the program model. Extensive quantitative and qualitative data formed a basis for presenting evaluation findings and conclusions pertaining to the four areas listed. This program theory-driven evaluation illustrates some of the common challenges involved in developing complex, innovative programs in a context where very little prior research or theory is available to guide program design. The strengths, challenges, and lessons learned from the theory-driven evaluation of the CIOF program are explored in detail in chapter 12.

7

Evaluation of the Health Insurance Policy Program

The purpose of this chapter is to illustrate how program theory-driven evaluation science was used to help design, implement, and evaluate a program to produce and disseminate knowledge of current statewide trends related to health and health insurance in the state of California. The Health Insurance Policy Program (HIPP) aimed to support efforts in the state to increase access to health insurance for California workers and their families. This chapter is organized to illustrate how the three general steps of program theory-driven evaluation science were implemented in this evaluation:

1. Developing program impact theory.
2. Formulating and prioritizing evaluation questions.
3. Answering evaluation questions.

The following presentation of the evaluation of the HIPP is intended to provide another specific example of program theory-driven evaluation in practice, and to show its applicability to programs outside the human services domain.

DEVELOPING A PROGRAM IMPACT THEORY FOR HIPP

Program Description and Rationale

With funding from The California Wellness Foundation (TCWF) to the University of California, Berkeley (UCB), a team mainly composed of researchers and staff from UCB and the subcontractor, the University of California, Los Angeles (UCLA) was formed to carry out the mission of the HIPP. The goal of the HIPP was to support the development of state policy to increase access to health insurance for employees and their dependents that was affordable, comprehensive, and promoted health and the prevention of disease. To this end, the HIPP strived to issue an annual report on the state of health insurance in California based on surveys of the nonelderly population, health maintenance organizations (HMOs), licensed health insurance carriers, purchasing groups, and employers. In addition, HIPP team members were expected to develop policy briefs and related health insurance publications for broad dissemination to appropriate policy stakeholders.

Program Impact Theory Development

During the first year of the project, the evaluation team began working with the grantees (UCB and UCLA project teams) and staff from TCWF to develop a program theory for the HIPP program. Discussions about program theory occurred over a 12-month period in face-to-face meetings, during conference calls, over e-mail, and through written document exchanges. The program theory that was used to guide the HIPP program and evaluation is shown in Figure 7.1.

The program theory shows that a range of publications development, report dissemination, and follow-up activities were conducted in an effort to increase target constituents' awareness and understanding of the status of health insurance issues in California and to influence policy development. Support activities and potential outcomes are shown in dotted line boxes to indicate that these were expected to occur but were not required by the funding agency. This representation of program theory, using rectangles with bullets and the dotted-line rectangle, was strongly preferred over other representations (e.g., circles and individual arrows like the previous two cases) by the stakeholders. Therefore, we used this format for the HIPP program theory, which turned out to be a useful guide for discussions of the HIPP evaluation questions.

FORMULATING AND PRIORITIZING EVALUATION QUESTIONS

The discussion of evaluation questions was also participatory and highly interactive. In collaboration with TCWF, the evaluation team led the discussions of potential evaluation questions with the two project teams. After exploring a wide range of possible evaluation questions, a

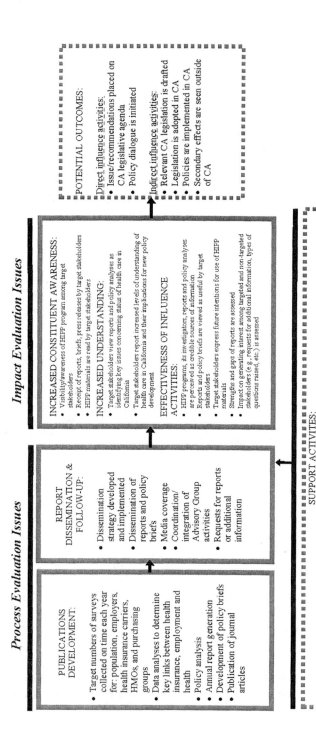

Figure 7.1. HIPP impact theory.

discussion of which questions were most important to answer, considering resource constraints and feasibility issues, was facilitated by the evaluation team. The group eventually decided to focus the HIPP evaluation on four key areas, including:

1. *Publications Development.* Did HIPP conduct surveys of California individuals, HMOs, insurers, employers, and purchasing groups? Were data analyzed and policy analysis conducted on the basis of data gathered?

2. *Report Dissemination and Follow-Up.* Were annual reports and policy briefs on the state of health insurance in California developed and disseminated to relevant policymakers?

3. *Research and Policy Recommendations.* Did HIPP identify key trends and make relevant policy recommendations regarding California health care needs, health status, risk status, and health risk behaviors? Did HIPP identify access barriers to affordable health care, health promotion, and disease prevention? Were findings from different surveys integrated to impact policy development and target outcomes in meaningful ways?

4. *Research Outcomes.* In what ways was HIPP effective in raising awareness among policymakers and the public about the status of health insurance in California and influencing the direction of health care policy?

ANSWERING EVALUATION QUESTIONS

This section provides a rather detailed account of the third step in the program theory-driven evaluation science process. To help readers understand how Step 3 can unfold in practice, data collection, key evaluation findings, and considerations for future programming for the HIPP program are now presented.

Data Collection

Interviews were conducted each year with a stratified, random sample of target audiences to address the evaluation questions. Key informants were stratified by organization type, including California policymakers, health insurers, HMO's, interest and advocacy groups, foundations, media, and university constituents. In-depth qualitative and quantitative data were gathered through face-to-face or telephone interviews to assess respondents' awareness of the research, understanding of material, how informative it was, and the influence of the research (i.e., credibility and usefulness). The evaluation team also examined whether and how respondents used the HIPP research in their work. In addition, media coverage and direct and indirect policy changes were tracked. Finally, a random sample of individuals publishing in the health insurance arena over the past 5 years was selected to provide a critical peer review of the HIPP research.

Overall, extensive qualitative and quantitative data were collected from over 300 key health care constituents over the program period, including three external peer reviewers. To supplement this, the evaluation team analyses also included a review of each research report and policy alert produced. Semiannual progress reports were also analyzed for key accomplishments and lessons learned.

To support continuous program improvement within the HIPP research program, the evaluation team prepared and disseminated 13 evaluation reports to HIPP program leaders over the 4-year funding period. These included four year-end evaluation reports, four midyear evaluation reports, and five interim reports. As tools for program monitoring and improvement, these reports documented not only key accomplishments and program activities, but also key program challenges and recommendations for addressing challenges. Conference calls and/or face-to-face meetings were held with program managers and TCWF program officers to discuss each report. During these various communications, the evaluation team facilitated discussion on the meaning of the findings and on developing strategies and responses to address recommendations. In addition, the evaluation team attended annual advisory group meetings and addressed questions pertaining to evaluation findings at those meetings.

Key Evaluation Findings and Conclusions

Publications Development. HIPP conducted annual surveys of California HMOs, health insurers, employers, and purchasing groups, and analyzed existing data sets of population-based surveys for 4 of the 5 years of TCWF funding. In the fifth year, 2000, the original program grant was terminated by the project's principal investigator. Although the surveys were conducted in the fifth year under a new contract, the evaluation team was not asked to evaluate the work performed under the new contract. As a result, only four out of five proposed annual reports, and 12 out of 20 proposed policy alerts were produced by the original program grantees and evaluated by the evaluation team.

Comprehensive data from multiple sources were collected during the first 4 years of funding. Four out of five targeted annual reports and 12 out of 20 targeted policy alerts were produced based on these data.

The number of reports produced increased from 1,513 in the first year to more than 2,500 in the fourth year.

With input from their advisory board, the HIPP decided to delay dissemination of the reports from October to January in order to capture the attention of the new California legislature each year. Due to budget issues and overruns in the costs of publishing the first annual report, TCWF authorized the production and dissemination of a stand-alone, six-page executive summary in lieu of two of the four policy alerts in the first year. Although the executive summary was produced and disseminated, the other two policy alerts were not developed. In years 2 through 4, the HIPP developed four two-page policy alerts each year. Although they were funded to produce four-page policy alerts, TCWF gave approval to the HIPP research team to produce two-page policy alerts starting in the second year of funding.

Overall, data collection efforts and survey response rates for the various sampling groups were commendable. Because of the policy advocacy uses of data that were collected, it was challenging for HIPP researchers to maintain high response rates for the health plan surveys. In contrast, HIPP was able to obtain 100% participation among all purchasing groups. Low response rates for employer surveys appeared to be at least partially due to poor management of survey administration by subcontractors. The specific annual response rate for each group was as follows:

- Purchasing groups—100% each year.
- Health maintenance organizations (HMOs)—83% to 94%.
- Health insurers—74% to 96%.
- Employers—30% to 50%.
- Behavioral Risk Factor Surveillance System (BRFSS)—response rate not reported, only sample size.
- Current Population Survey (CPS)—response rate not reported, only sample size.

A challenge that emerged during the 1998 data collection efforts involved contract renegotiation with KPMG regarding the Employer Health Benefits Surveys. KPMG incurred significant cost overruns in their attempt to improve sampling efforts in gathering 1997 data, and subsequently requested a substantial increase in funding to conduct the employer surveys. This required UCLA to renegotiate the HIPP contract with KPMG at a substantially higher rate, and secure additional funding. To cover increased employer survey costs, TCWF gave HIPP an augmentation grant of $24,500 to cover years 2 and 3, and an additional $12,250 per year for years 4 and 5.

HIPP researchers produced four comprehensive annual reports. These reports consolidated and described key research findings in each area that was surveyed. Recipients of HIPP reports appreciated the comprehensiveness of the annual reports. There was consensus that the HIPP reports filled an information gap and provided the only place that such data could be found in one resource. Also, based on evaluation feedback, HIPP researchers expanded their analyses over time to include more detail focused on demographic differences and the geographic levels of analysis. Each report increased in length, from 90 pages in the first report to 131 pages in the final report.

Comments from peer reviewers of the HIPP reports suggested that determining the generalizability of the results was made difficult due to the lack of reporting on key sampling statistics. In their view, HIPP researchers did not clearly discuss the limitations of the data due to types of data collected and research methodologies employed. They cautioned that these types of omissions might potentially mislead "lay" readers into making generalizations that were not supported by the data. Because the evaluation team conducted a peer review as part of the summative evaluation, the HIPP did not have an opportunity to address this feedback through the continuous improvement process. For example:

- Although samples obtained in the Behavioral Risk Factor Surveillance System (BRFSS) and the Current Population Survey (CPS) reflected significant sample sizes, overall response rates and sampling methodologies were not reported in HIPP publications.
- HIPP researchers did not clarify how data were analyzed to take into account the nature of the data and sampling procedures.
- Limitations within the data and their implications for interpreting findings were not adequately discussed in HIPP research products.

More Policy Analysis Desired by Targeted Constituents. Annual feedback from respondents indicated that some did not understand how the HIPP researchers arrived at certain policy recommendations. More specifically, evidence was not presented concerning how the data informed many recommendations, or why recommendations were the best options. There was also a desire for a more comparative analysis of policy options as well as evidence-based solutions for specific populations in need. For instance, there was a need for more background and critical analysis of strengths and weaknesses of the recommendations offered.

Competition between researchers and role conflicts affected the collaboration between HIPP researchers and posed key challenges to the development of publications, particularly in the early phase of the program. The lack of agreement about roles and responsibilities for writing and approving HIPP reports and materials contributed to the internal conflicts between HIPP researchers. It appeared the HIPP researchers had different styles and beliefs about the way tasks should be accomplished. Both seemed reluctant at times to compromise.

Coordination and Integration of Advisory Information. The HIPP advisory group was convened each year in November to review the draft outline for each annual report and to review the findings and recommendations to be included in the reports. The purpose of these meetings was to get feedback on the findings and policy recommendations and obtain input on new issues to be considered in future reports. All data shared with advisory group members was confidential, and findings were embargoed until they were officially released to the press each year. The advisory group recommended, and the foundation approved, that the release of the initial report and each subsequent report would be delayed from the fall of each year to January of the next year in order to accommodate and capitalize on the attention of the new administration put into office.

Report Dissemination and Follow-Up. Approximately 7,000 reports were disseminated to targeted health care constituents over a 4-year period. The number of annual reports and policy alerts produced and disseminated increased each year, from 579 annual reports released in year 1 to 3,000 in year 4. The primary method for dissemination was mailing the reports to key constituents in health care, government, legislature, media, interest and advocacy organizations, universities, and private organizations. Other dissemination vehicles employed included distributing copies at conferences and roundtables, sending copies to individuals who requested them, and electronic dissemination. A summary of the HIPP dissemination efforts is provided in Table 7.1.

HIPP reports and research findings were also disseminated by HIPP project teams to diverse audiences through more than 70 presentations at conferences, roundtables, and invited addresses. In addition, the HIPP research teams gave several briefings and testimonials on the data to policy-oriented audiences as well as technical assistance to others addressing health insurance issues in California.

HIPP researchers also responded to numerous requests for additional policy analyses and data estimates each year. These included more than 200 requests each year from health care constituents, including state government agencies and officials, state legislative staff, academic, advocacy, and policy research organizations, private health care organizations, and others.

TABLE 7.1

Health Insurance Policy Program Dissemination Efforts

Report Dissemination and Follow-Up Activities	1996 Report	1997 Report	1998 Report	1999 Report
# Reports Produced	1,513	1,800	2,000	3,500
# Reports Disseminated	579	1,800	2,000	3,000
Release Date	January 1997	January 1998	January 1999	January 2000
# Policy Alerts Produced	0 (instead, produced a 6-page executive summary)	4	4	4
# Policy Alerts Disseminated	1,605 (executive summaries)	1,500 of each policy alert (Feb., Mar., Apr., May)	1,700 Feb. & Mar. Alerts 2,000 Apr. & May Alerts	1,915 Feb.; 1,760-Mar., 2,000-Apr., 1760 May
Electronic Dissemination	—	More than 80 copies of the 1997 report were downloaded	More than 320 copies of the 1998 report were downloaded	More than 200 copies of the 1999 report were downloaded
Dissemination Audiences	Mailing List: health care, government, legislature, university, media, and private organizations	Mailing List: health care, government, legislature, university, media, and private organizations	Mailing List: conferences, roundtables governmental agencies, nongovt. agencies, and requested copies	Mailing List: conferences, roundtables governmental agencies, nongovt. agencies, and requested copies
Requests for additional policy analyses and/or data estimates	200	213	226	174
Supporting Activities				
Presentations	4	18	37	15
Papers	—	—	9	8
Briefings/Tech. Asst./Testimonials	—	8	30	4

Dissemination Strategy. The HIPP program had a weak dissemination strategy in the first year, but significantly improved dissemination efforts over time with the aid of a media and communications consultant. Early dissemination problems included lack of a clear dissemination strategy, lack of a clearly defined target audience, poor contact information for target constituents, and lack of coordination of dissemination efforts, particularly around dissemination of press releases.

Dissemination Targets. HIPP researchers had a difficult time identifying, developing, and sharing an agreed on list of target constituents in the first year. Follow-up efforts to clarify target audiences revealed to the evaluation team that the investigators did not know the organizational affiliation of most of the target constituents to whom they sent their first report. In addition, more than half of respondents had insufficient or inaccurate contact information. The evaluation team worked collaboratively with HIPP researchers at UCB to locate and update contact information in the first year. Subsequently, the HIPP team at UCB worked to update contact list information each year.

Initial Dissemination Strategy. HIPP researchers did not have a well-defined dissemination strategy developed during the early phase of their program. Both institutions relied on their respective public relations departments but did not adequately coordinate their efforts. Key factors that were lacking included (a) a clearly defined dissemination audience, (b) a clearly defined approach for working with print and broadcast media representatives, and (c) a dissemination plan that included coordinated production and release dates.

Technical Assistance With Dissemination Efforts Were Needed. To assist in strengthening dissemination efforts, TCWF supplied additional funding to the HIPP in the amount of $10,000 per year in years 3, 4, and 5 to cover costs of retaining a media consultant. Input from their media and communications consultant in subsequent years proved very helpful in clarifying target audiences and developing a clear dissemination strategy that included coordination of production and dissemination schedules and press releases. The consultant was particularly helpful in assisting HIPP researchers make initial media contacts and prepare talking points for speaking with the media. Specific efforts were also made to reach non-English speaking language media outlets.

Coordination Challenges. Miscommunication between HIPP researchers resulted in failure to coordinate the release of year 1 findings by their respective institutional public relations offices. This was attributed by the project's principal investigator as a cause for lack of coverage in key news outlets such as the *LA Times* in 1997. As a result, responsibility for report production and dissemination was shifted from UCLA to UCB in year 2. In spite of establishing mutually agreed-on dissemination schedules, further coordination problems were experienced in 1998 and 1999.

Media Coverage. The HIPP received media coverage each year. Coordination challenges resulted in few stories covered in northern California and no print coverage in southern California in 1996. Coverage increased dramatically, however, with the release of the second publication. The HIPP team was successful at getting coverage in the state's largest paper, *The LA Times*, in the fourth year of dissemination, and coverage in the *San Francisco Chronicle* for 3 of the 4 years in which findings were released. With input and direction from their media consultant, the HIPP was able to gain coverage over key wire services in years 2 through 4. This resulted in significantly greater coverage in both broadcast and print media. Table 7.2 summarizes key types of media coverage obtained throughout the funding period.

Research and Policy Recommendations

HIPP examined several core health insurance topics over time:

- Health insurance coverage of Californians.
- Commercial health plans in the California market.
- Medi-Cal Managed Care.
- Purchasing groups.
- Employer's experiences and practices in providing health benefits to their employees.
- The integration of health promotion and public health into California's health care system.

Data were analyzed to determine the key links between health insurance, employment, and health. The key trends identified were the primacy of employer-provided health insurance in California and the poorer health of the uninsured. The key links between health insurance, employment, and health include, but are not limited to:

- Most Californians get their insurance through employer-provided insurance.
- Most uninsured Californians are working families and individuals, including 6 in 10 who are full-time employees or their dependents.

TABLE 7.2
Summary of HIPP Media Coverage

Coverage Type	1996 Report	1997 Report	1998 Report	1999 Report
Print Media				
Major Print Outlets	6 publications	25 publications	20 publications	25 publications
	SF Chronicle	Boston Globe	SF Chronicle	Los Angeles Times
	Contra Costa Times	Wall Street Journal	Sacramento Bee	Wall Street Journal
			San Jose Mercury News	Sacramento Bee
			Oakland Tribune	SF Chronicle
			San Diego Union	San Jose Mercury News
			Orange County Register	Oakland Tribune
			La Opinion	San Diego Union
				Orange County Register
				La Opinion
Broadcast Media Statewide	1 Broadcast Source	10 Broadcast Sources	7 Broadcast Sources	6 Radio Stations TV Coverage
Major Broadcast Outlets	KCBS Radio	Bay TV	NBC News Affiliate in Sacramento, Univision, Ch. 34 CBS Network News Radio in NYC	KFWB (Hollywood), KQED (SF), KCSN (LA), KPIX (SF)
Wire & Electronic Services	0	3	3	5
Wire Services	—	Associated Press CNN Business Wire	Associated Press CNN Business Wire	Associated Press CNN Business Wire
Electronic Outlets	—	—	—	California Healthline Health Care Daily On-line
Media Contacts	Not tracked	200	120	50+

- More than half of uninsured adults cite job-related reasons for their lack of insurance coverage.
- Uninsured adults ages 18 through 64 in California have poorer health status, higher rates of preventable health risks, and less access to preventive care than do insured adults.
- Two thirds of all uninsured Californians have low family incomes and two thirds of uninsured adult employees make less than $15,000 a year.

The HIPP identified several key barriers to affordable health care, health promotion, and disease prevention. The core barriers include lack of access, cost, limited choice, and cost-coverage tradeoffs. Other factors include lack of standards, enforcement, and accountability among health care providers. According to HIPP findings, specific key barriers to affordable health care, health promotion, and disease prevention include, but are not limited to:

- Affordability for working families.
- Affordability for employers, particularly small and mid-sized firms.
- Lack of agreed-on standards for minimum health care and health insurance coverage.
- Lack of health plan accountability for denials and delay of needed care.
- Lack of knowledge about or access to health insurance purchasing groups.
- Low utilization of health promotion programs in health plans.
- Low offer rates of health insurance by employers.
- Lower insurance eligibility and take-up rates for certain groups of individuals.
- Few employers offer worksite wellness programs to their employees.

Numerous policy recommendations were made in each annual report and subsequent policy alerts produced by HIPP. Collectively, over 100 policy recommendations were offered in several areas to address various ways to increase access to affordable, quality health insurance for the uninsured in California. The core areas that policy recommendations addressed included:

- State and federal health insurance reforms.
- Managed care and other health plan reforms.
- Medi-Cal managed care.

- Purchasing groups.
- Strengthening public health programs and health promotion policy.
- Healthy families and the uninsured.

A number of policy recommendations were restated in more than one report. Among these, key recommendations included, but were not limited to:

- Standardizing benefit packages and eliminating individual mandatory contributions.
- Guaranteed issue and renewal of health insurance for low-wage and/or low-profit small employers.
- Developing, monitoring, and enforcing health plan standards, requirements, and accountability.
- Ensuring access to direct-care services to the uninsured.
- Establishing a statewide purchasing pool for mid-size businesses.
- Providing a voice for consumers.
- Creating new review, grievance, enforcement, and reporting entities.
- Creating a comprehensive, publicly available, statewide, automated state health data system on California's managed health care plans and network providers.

Research Outcomes

Overall, responses regarding the reports' effectiveness, impact, credibility, and usefulness were very positive. Compared to other topics covered in reports, sections dealing with health insurance coverage and demographic breakdowns of the uninsured were perceived to be most informative, and were most likely to be viewed by respondents as the key messages being presented. In general, a higher percentage of respondents found the full report (in comparison to the policy alerts) to be "very useful" and "very effective" at capturing their interest and attention. In light of HIPP's goal of increasing access to health promotion and disease prevention in health insurance, respondents were least likely to find sections covering these issues to be informative. In addition, no respondents identified these as the key messages in the HIPP reports.

The impact of HIPP research reports on increasing constituent awareness and understanding of the state of health insurance in California, as well as its impact on influencing policy-oriented outcomes, was assessed through a series of annual, structured interviews. With input from and agreement by HIPP program leaders and TCWF program officers, the evaluation team conducted follow-up interviews during the first 3 years of report production with random samples of key informants who were targeted for dissemination and influence. To ensure that the evaluation team obtained feedback from a broad range of constituents, they first stratified each list of report recipients according to organizational affiliation (e.g., state policy makers, health insurers, media, etc.) before randomly selecting names within each stratum to be interviewed. Confidential interviews conducted in year 3 were drawn from a list of key informants who were highly involved in the health insurance policy arena. Interviews were also conducted with a random sample of researchers who published in the health insurance arena but who were not directly connected with the HIPP program. These subject-matter experts provided an additional critical peer review of the HIPP research products.

Response Rates. Overall response rates for follow-up interview efforts ranged from 47% in the first year to 66.7% in the third year. The main reason for the low response rate in year 1 was that most respondents had insufficient contact information, and could not be reached. The key reason for nonresponse in years 2 and 3 was that respondents were too busy to participate. After comparing characteristics of those who were and were not contacted during follow-up time periods, very few significant differences were detected, suggesting that individuals reached were generally representative of those who were targeted. However, data must be interpreted in light of potential problems posed by imperfect response rates.

Awareness

Awareness of the HIPP reports and research program among target constituents increased from 63% in year 1 to 100% in year 3. Most respondents learned about HIPP from receiving copies of HIPP reports and policy alerts. Others learned about HIPP through personal contact with HIPP researchers or at professional conferences. Available evidence suggests that improvement in dissemination efforts was crucial for increasing awareness. Lower levels of awareness were related to higher numbers of constituents who did not receive the annual reports or policy alerts, as well as an increase in the proportion who could not recall whether they received HIPP reports. In addition, increased awareness was related to an increase in the proportion that read at least some of the research reports. Compared to 15.2% of respondents who did not read any of the first annual report, only 6% of respondents did not read any of the third annual report. Overall, more respondents recalled having received and being familiar with the annual reports than the policy alerts.

By examining the level of interest and attention generated by the reports and policy alerts, the awareness of the respondents was assessed. Toward this goal, respondents were asked each year to rate how effective the reports were in capturing their interest and attention. Overall, between 90% and 100% of all respondents and between 84% and 100% of policymakers reported that each of the annual reports was "somewhat" or "very effective" in capturing their interest and attention. Although most respondents also felt the policy alerts were effective, in general, respondents were more likely to view the full report as effective in capturing their interest and attention.

Understanding

By asking respondents to describe the key messages received from the reports and to rate what they felt was most informative in each report, the understanding of the respondents was assessed.

Most Informative Sections. Nearly all health care leaders surveyed each year felt that all sections of the HIPP reports were informative. The sections on health insurance coverage were most consistently rated as informative over time. Similarly, respondents were much more likely to recall sections on health insurance coverage than other health care issues discussed in the reports.

Key Messages Taken Away. Compared to the full reports, respondents were less likely to recall receiving the policy alerts, and less likely to recall what information was contained in them. When asked to relay the key findings of the reports in their own words, the most frequently recalled findings concerned the demographics and numbers of uninsured in California. Other issues recalled included the general state of health insurance coverage, health insurance problems in California, and the "growing" problem of numbers getting higher and issues not being satisfied. No respondents recalled or identified key issues related to the integration of health promotion and public health into California's health care system.

Influence

By examining the perceived credibility and usefulness of each report and policy alert, the influence of HIPP research reports was assessed. In addition, the evaluation team also examined whether or not and how respondents were actually using the information from the reports in their daily work.

Perceived Credibility of Reports.   More than 9 in 10 respondents felt that the annual HIPP reports provided accurate representations of the state of health insurance in California. Compared to just over half of all respondents in 1996, nearly three fourths of respondents in 1997 and 1998 felt the reports were "very accurate."

Usefulness of HIPP Research.   The majority of respondents found all sections of the HIPP reports and policy alerts to be useful. California policymakers found the sections on health insurance coverage and health promotion particularly useful to them. Compared to the policy alerts, respondents were more likely to say that the policy recommendations in the full report were "very useful." Respondents were also more likely to criticize the policy alerts and policy recommendations than the full report and research findings. Furthermore, among those who had on occasion requested additional data estimates or policy analyses from either of the HIPP researchers all said that they were "very" or "somewhat useful" to them.

How HIPP Reports Were Used by Respondents.   Nearly all respondents said they used the HIPP research in some fashion. When asked about how they used the reports and policy alerts in their work, half of all respondents specifically cited using the full report, with the remaining mentioning how they used the numbers or information more generally. Only one respondent specifically mentioned using the policy alerts (which were used in developing policy positions and drafting legislation). Primarily, respondents used the HIPP research as reference material for their written publications, including articles, fact sheets, letters, and briefings. Additionally, respondents reported that they referenced HIPP research for speeches, public speaking engagements, interviews, presentations, and testimonies. Many also used the reports as general reference material. Approximately one fifth said they used the material to educate their staff, policymakers, and other organizations, and nearly as many said they used the statistics provided in the report in their work. Only a handful of respondents reported using the HIPP research in their policy development efforts.

Recommendations for Improvement.   Many respondents had no recommendations for improvement. Many said they liked the reports and felt they were well-written and organized, and provided useful, important information. Among recommendations offered, key suggestions focused on the packaging of the report. Several comments called for increasing the length and detail of the reports, and including more depth of analysis, including adding personal stories as part of the report to connect with readers on a more personal level. Other feedback involved requests for more comparative data (e.g., comparative analysis with the nation and non U.S. types of public coverage and types of funding) and increased and/or continued use of charts and graphs. Comments pertaining to the policy alerts included requests for additional supporting data about how conclusions were drawn.

CONSIDERATIONS FOR FUTURE PROGRAMMING

This section briefly highlights key information or lessons learned that might help improve future programming by TCWF in this area. Future efforts similar to the HIPP should take into account that (a) comprehensive reports were more effective than policy alerts, (b) media and communications technical assistance was crucial, and (c) productive research collaborations across universities were difficult to sustain.

Comprehensive Reports Were More Effective Than Policy Alerts. Compared to the policy alerts, the available evidence suggests that health care constituents are more likely to value and use comprehensive data in a single source such as the HIPP annual report. Respondents were more likely to recall receiving these reports and they were more likely to recall information provided in them. They were also more likely to rate the policy recommendations in the report as useful and informative than those in the policy alerts. Of particular importance is that sufficient data are presented and analyzed in a manner to justify policy recommendations.

Media and Communications Technical Assistance Was Crucial. The ability to conduct quality research does not guarantee that researchers have strong skills in advocacy and dissemination. University public relations offices are typically unreliable when asked to support projects of this nature. The additional resources made available to hire a media and communications consultant added tremendous value in helping HIPP researchers (a) clarify their target audiences, (b) develop strategies for working with the media, (c) develop talking points, and (d) increase media exposure and coverage.

Productive Research Collaborations Across Universities Were Difficult to Sustain. TCWF created the formal partnership between UCB and UCLA researchers. Although there was general consensus among HIPP stakeholders, including UCB and UCLA researchers, that the HIPP was able to accomplish more because of the combined efforts, it was very difficult to sustain this type of collaborative arrangement over time. Future efforts of this sort may benefit from (a) well-designed processes to carefully select the partners, (b) a process to facilitate ongoing role clarification and communication, (c) a process to manage conflict, and (d) a process to develop and reward teamwork.

CONCLUSION

The application of program theory-driven evaluation science to improving and evaluating a rather complex research and policy influence program was presented in this chapter. A nonconventional program theory was used to focus this program theory-driven evaluation on questions that fell within four main areas: publications development; report dissemination and follow-up; research and policy recommendations; and research outcomes. In-depth qualitative and quantitative data were gathered to form the basis for answering the evaluation questions deemed most important by the HIPP project teams. Evaluation findings and evaluative conclusions were presented to illustrate sample products from the theory-driven evaluation of the HIPP program. The strengths, challenges, and lessons learned from the theory-driven evaluation of the HIPP are explored in detail in chapter 12.

8

Evaluation of the Future of Work and Health Program

The purpose of this chapter is to illustrate how program theory-driven evaluation science was used to help design, implement, and evaluate a program to understand the rapidly changing nature of work and its effects on the health and well-being of Californians. This chapter is organized to address the three steps of program theory-driven evaluation science:

1. Developing program impact theory.
2. Formulating and prioritizing evaluation questions.
3. Answering evaluation questions.

The following presentation of the evaluation of the Future of Work and Health (FWH) Program is intended to illustrate yet another example of program theory-driven evaluation science in action.

DEVELOPING A PROGRAM IMPACT THEORY FOR THE FWH PROGRAM

Program Description and Rationale

The FWH program aimed to understand the rapidly changing nature of work and its effects on the health of Californians. This program was designed to support a range of research projects and statewide meetings consistent with this aim. Key program objectives included (a) identifying issues and trends important to the future of work and health of Californians, (b) building a network of people involved in expanding knowledge and improving practice to advance the future of work and health in California, (c) funding projects to illuminate trends important to the future of work and health of Californians, (d) identifying policies that can influence work and health trends to improve the health of Californians, and (e) disseminating research findings on California work and health trends from the FWH program activities.

Modification of Original Program Goals. The FWH program goals changed substantially from program inception. The primary mission of the program changed from (a) improving health and well-being of Californians through employment-related approaches targeting policy changes, organizational practices, and research to (b) increasing our understanding of the changing nature of work and its impact on the health of Californians through research and discussion.

The original mission of FWH focused on improving the health and well-being of Californians by fostering improvements in work-related organizational practices and public policies that affect the health of workers, their families, and their communities. The means for achieving this was to engage researchers and convene leaders from business, government, labor, academe, and the media to examine (a) the changing nature of work and devise ways to improve preparation for work, (b) terms of employment, (c) conditions of work that foster good health and productivity, (d) access to employment and the overall climate for employment, and (e) wage growth in California.

Without a clear guiding program framework, the original FWH principal investigator continually reworked program goals during 1996 toward a revised mission that centered on workforce preparation and training opportunities. More specifically, the FWH program revised its goals to move in the direction of encouraging better linkages between employment-based opportunities for workforce preparation and the educational system, as well as recommending organizational practices and public policies that can help expand employment-based opportunities and build educational system capacities.

During 1996, the evaluation team engaged the FWH program leader in several discussions and meetings about how the program could be structured to achieve maximum success. The focus of these communications included, but was not limited to, clarifying program objectives, understanding procedures and criteria used to select program projects, and identifying the kinds of data that could be collected to address these issues. In the process of discussing how the evaluation would be focused, the proposed goals of the FWH program were further revised, and repackaged into four overarching objectives:

1. *Research.* The FWH research objective was to integrate usable, existing knowledge and, where possible, to create new knowledge

about how better workplace preparation and practices lead to better worker health and well-being, as well as increased worker productivity. In order to achieve this goal, the FWH strategy was stated as: enlisting thought leaders to write perspective pieces covering how work and health are considered by various disciplines, and generating interest in other foundations in underwriting this initial exploration and in building this new field of research.

2. *Demand Creation.* The aim of the "demand creation" goal was to send messages about better work preparation and practices that can change behavior on the part of individuals, employers, and public officials. Toward this goal, a two-part strategy was to first develop or synthesize compelling information about work and health in the context of everyday people's lives, and second, to package and promote that information so that it becomes readily accessible to Californians.

3. *Catalyzing Change.* The objective of "catalyzing change" was to induce changes in individual behavior, organizational practices, and public policies and to develop groups to act on TCWF recommendations. The strategy for catalyzing change was to develop ownership and responsibility among the organizations involved early on in the agenda-setting process.

4. *Develop Action–Research Infrastructure.* The objective here was to continue to increase the amount of research and research-based action on work and health past the life of the program. To meet this objective, TFWH planned to build a research and action network by (a) identifying the right participants, (b) introducing these people to each other and helping them gain familiarity with each other's work, (c) establishing groups of individuals around common interests and projects on which they can work together, (d) creating collaborative projects, (e) catalyzing joint action and establishing their "ownership" of both the process and the work it generates, and (f) establishing a structure of information associations composed of organizational directors who meet quarterly to facilitate collaborative efforts.

Key Challenges with New Overarching Objectives. Key challenges faced in year 1 stemmed from a lack of an underlying conceptual framework or program model that (a) linked the objectives and strategies together, and (b) focused the content and activities of the program. Additional challenges evident under the original program structure included:

- Ensuring that the most important links between workplace changes and health were identified, explored, and made explicit within each funded activity.
- Providing a framework or rationale for the range of trends chosen for examination.
- Ensuring that specific activities were designed to meet action-research objectives.
- Establishing some mechanism and/or guiding criteria for identifying and including the most relevant stakeholder organizations throughout California to address work and health issues.
- Ensuring that the combined efforts of each of the funded projects and activities were representative of the range of work and health issues experienced by the many diverse populations in California.
- Developing a clear strategy for identifying best organizational policies and practices beyond workforce training and preparation (e.g., worksite health promotion) to ensure organizational and policy level change.
- Identifying realistic outcomes that can be expected throughout the duration of the grant.
- Identifying and building constituent support for and involvement in program activities within the time frame of the funding.

To address these challenges and improve program design, the evaluation team recommended that the program could be strengthened by developing a framework grounded in a review of the work and health research literature, current corporate practices and public policies, and trends in the workforce.

FWH Program Was Reconfigured Again.   The original grantee decided to terminate its grant on completion of its first program year. Responsibility for the redesign of the program, as well as for program management, shifted internally to TCWF. Program goals were revisited and the program was reconfigured during 1997 and 1998 to address many of the challenges identified by the evaluation team in year 1.

One of the first changes concerned a change in program management. Whereas the original program was led by an external grantee, leadership for the revised program was taken over by the senior program officer for the WHI. Several changes were also made to the focus and structure of the program to enable the program to be completed within the remaining budgetary and time-frame constraints of the initiative. Most notably, the mission of the program was modified so that the aim of the revised FWH program was to understand the rapidly changing nature of work and its effects on the health of Californians. To accomplish this, the structure of the program was redesigned to support a range of research projects and statewide meetings consistent with this aim. In addition, the roles of evaluation and continuous improvement were reduced so that funded projects under the new structure would not have an external evaluation component.

To assist in developing a framework for the revised FWH program and to assist in program coordination, TCWF funded the Institute of Regional and Urban Studies (IRUS) in June of 1997. Specifically, IRUS was funded to work closely with TCWF in the development, implementation, and dissemination of FWH funded programs. Together, IRUS and TCWF convened a panel of work, health, and economic experts to identify those issues and trends most important to the future of work and health in California. These trends, in turn, served as a framework for identifying and funding research and convening projects that would illuminate key work and health linkages in these areas. By April 1998, the panel reached agreement on three broad trends. The trends were (a) getting left behind in a changing economy, (b) a widening of income inequality, and (c) a changing contract between employer and employees.

Several research and convening projects were then funded throughout 1998 to address the three broad trends identified by the FWH panel. In March 1998, the Board approved a grant recommendation to the University of California, San Francisco's Institute for Health Policy Studies (UCSF IHPS) to conduct a 3-year longitudinal "California Work and Health Survey" of a representative sample of Californians to better understand important work and health connections. IHPS was also funded to convene and support additional research on the longitudinal dataset. In addition, the Board approved eight grants selected from 64 proposals reviewed for additional research and meetings totaling $865,000 in December 1998 to address the trends identified through the panel process. See Table 8.1.

The evaluation team held numerous meetings and conference calls with TCWF and IRUS to clarify the goals of programs and evaluations. Developing a program theory for the revised program was quite challenging for a number of reasons. First and foremost, the evaluation team was put into the awkward position of having to evaluate the work of the initiative's senior program officer who assumed a new program management role. It was difficult for the evaluation team because the program officer had two conflicting roles in relation to the evaluation, and the evaluation team had already established a very different type of working relationship with the officer as the TCWF leader of the entire initiative.

TABLE 8.1
Organizations Funded to Develop FWH Research and Convening Programs

California Work and Health Survey Longitudinal Research:

- University of California, San Francisco: Institute for Health Policy Studies

Additional Research and Convening Grantees to Address the 3 Themes:

• American Institutes for Research in the Behavioral Sciences	• Industrial Areas Foundation (IAF)
• California Center for Health Improvement (CCHI)	• Kaiser Foundation Research Institute (KFRI)
• California Institute for Rural Health Management (CIRHM)	• UC San Francisco Institute for Health & Aging (IHA)
• FAME Assistance Corporation	• UCLA Institute of Industrial Relations

Program Impact Theory Development

Identifying what could be evaluated required the program to be reconstituted again and projects put into place. It was agreed that work could not begin until after the new projects were selected. However, once programs were funded, the evaluation team learned that the new grantees were told that they were not expected to work with the evaluation team. Instead, they were informed that TCWF would maintain direct responsibility for evaluating program efforts. The conflicting role of having to evaluate the FWH program without having access to program grantees limited what could be accomplished by the evaluation team. In addition, the evaluation team experienced considerable difficulty gaining the cooperation of key representatives of the FWH program in the evaluation process. In order to clarify the new evaluation mandate for the revised FWH program, the evaluation team eventually gained agreement with the foundation that the focus of the evaluation would center on the overall program as a unit of analysis, rather than on specific grantees or projects within the overall program. Figure 8.1 illustrates the program theory that the evaluation team ultimately developed to guide the FWH program evaluation.

FORMULATING AND PRIORITIZING EVALUATION QUESTIONS

Although the discussions of program theory and the evaluation questions were participatory and highly interactive, it was challenging to reach consensus on what and how to evaluate the FWH program. After much discussion and a fair amount of conflict, the relevant stakeholders agreed to address the following questions:

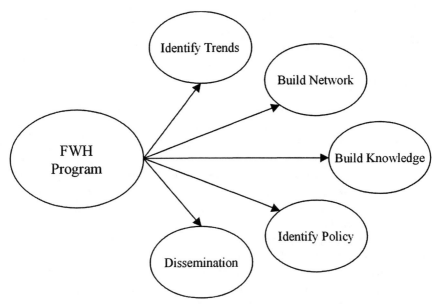

Figure 8.1.   FWH program impact theory.

1. *Trends.* How effective was the FWH research and meetings program in identifying important trends about the future of work and health in California? What trends were identified?
2. *Build a Network.* How effective was the FWH program in creating a network of individuals working together on work and health issues? Who was involved in the network? What did they gain from their collaboration?
3. *Build Knowledge.* How effective was the FWH program in building knowledge about work and health connections in California? How well did program efforts help others understand the links between work and health?
4. *Identify Policy.* What policy-suggestive findings were identified that can influence work and health trends and improve the health of Californians?
5. *Dissemination.* How were FWH research findings disseminated? How informative and useful were these to target constituents?

ANSWERING EVALUATION QUESTIONS

This section provides a rather detailed account of the third step in the program theory-driven evaluation science process. To help readers understand how Step 3 can unfold in practice, data collection, key evaluation findings, and considerations for future programming for the FWH program are presented.

Data Collection

The FWH evaluation focused on the effectiveness of overall program efforts in five core areas targeted in the program: identifying trends, building a network, building knowledge, identifying policy, and dissemination. More specifically, it was agreed that the evaluation team would take a historical and descriptive approach to evaluating the effectiveness of FWH program efforts in reaching these goals. It was also agreed that the evaluation team would conduct follow-up interviews with a random sample of individuals who participated in the culminating FWH meeting to assess the effectiveness and usefulness of the conference, as well as knowledge gained.

Toward these goals, the evaluation team conducted an extensive document analysis review of more than 150 documents produced through the FWH program. The core documents included semiannual grantee progress reports, research reports, and publications. Also included were reviews of annual press releases and press coverage, internal memos, and notes and summaries from program meetings. All documents were analyzed for key accomplishments and lessons learned. A database was created to assist tracking and analyzing program goals, activities, and key issues and trends identified in the documents. Finally, follow-up telephone interviews were conducted with a stratified random sample of 36 individuals who attended the final program meeting in March 2001. Respondents were stratified according to whether they were internal (i.e., individuals who were funded as part of TCWF WHI) or external (not part of the initiative) constituents. Prior to sampling, the evaluation team excluded members from the three foundations who funded the conference as well as members from the evaluation team to ensure the interviews focused on target constituents rather than on conference sponsors.

Under the reconstituted FWH program, the new grants did not include using evaluation to help grantees improve their performance. Thus, evaluation efforts were shifted from providing continuous program improvement feedback (as was done in year 1) to conducting a summative evaluation at the completion of program activities in 2000. In light of this, the evaluation team prepared and disseminated only two evaluation reports to FWH program leaders over the funding period. The first included an evaluation review and an interim evaluation report disseminated during the first program year.

Key Evaluation Findings and Conclusions

Identifying Trends. As mentioned previously, three key trends were identified and focused on by the FWH program. The trends were: (a) getting left behind by a changing economy, (b) a widening of income inequality, and (c) a changing contract between employer and employees. However, findings from many of the research projects funded by FWH program did not seem to directly address the key trends.

Summarizing all of the key findings from each research project is not within the scope of this chapter. Rather, the goal is to characterize the type of research that was conducted and to highlight some of the key findings pertaining to the work–health nexus. These are explained in the following discussion of research efforts, types of issues examined, and key findings.

Research Projects. A number of research efforts were funded under the FWH program to identify work and health trends in California. The two largest data collection efforts included the 1996 survey, "The Work and Health of Californians," conducted by The Institute for the Future, and the longitudinal "California Work and Health Survey (1998–2000)," conducted by UCSF IHPS. Together, these data collection efforts examined both the employment status and health status of Californians, and their relationship to one another using representative samples of adult Californians ages 18 to 64. Included in these datasets is extensive information about the nature of Californians' employment situations, personal and family backgrounds, and status with respect to physical and mental health. In addition, eight projects were funded to address the three broad themes of the FWH program. The research topics addressed by these programs included: "Leading California Companies," "Work and Health Public Policies," "Rural California," "The Role of Churches," "Building Social Capital," "Income Inequality," "Older Workers," and "Organized Labor." Finally, to stimulate further research on work and health trends in California, UCSF researchers awarded $5,000 research subcontracts to several individuals each year during the 1998–2000 period for conducting secondary data analyses of the longitudinal "California Work and Health Survey" data set.

Issues and Trends Examined. A vast number of variables and issues were explored across the various research efforts that took place within the FWH program. The majority of analyses conducted did not examine trends over time. Rather, most examined the types of relationships that existed between various employment and health related variables at one point in time. All of the research examined various relationships between work and health issues. It is our opinion, however, that the FWH program did not synthesize these diverse findings in a manner that led to identifying the issues and trends most important to the work and health of Californians. Most efforts focused on disseminating key findings from each project through collections of papers and proceedings, through press releases to the media, or through handouts that were shared at grantee meetings. Although numerous meetings took place to discuss specific work and health findings from each project, these efforts did not include mechanisms for identifying and prioritizing which of the key findings had the most important work and health implications for Californians. Further, a key limitation was that most projects did not clearly inform one or more of the three key trends identified through the panel process. Only a fraction of all projects appeared to map directly into one or more of the three program themes.

Collectively, more than 150 different variables were included in analyses of data collected and analyzed in the various datasets. Approximately one fifth of FWH-funded projects consisted of new data collection efforts, and more than half consisted of projects that conducted secondary data analyses of existing data sets or data sets collected within the FWH program. The variables studied most frequently included health status, employment status, health benefits and/or health insurance, race and ethnicity, income, age, gender, mental health, education, work hours, poverty, presence of children in the home, job loss, marital status, socioeconomic status, stress, union membership, skills for work, chronic illness, disability, cigarette smoking, retirement, prosperity, and work schedule.

Other variables examined in several efforts included factors such as social capital, temporary and/or contingent workers, availability of child care, language, employment and/or promotion opportunities, firm size, decision autonomy at work, health-risk behaviors, and workplace community. In addition, a handful of studies examined the following topics: anticipated hardships, college education, skilled workforce, death rate, life expectancy, computer access and literacy, social networks, community well-being, culture, industry type, organizational outcomes, job demands, working conditions, household composition, supervisory authority, social security, private pension, job characteristics, immigrant status, flexible work schedules, perceived control, physical job demands, job tenure, and difficulty living on household income.

Research Topics Explored Within the Longitudinal CWHS Dataset.
Topics examined by research grantees in 1998 included "Employment Discrimination and Its Health and Mental Health Correlates Among Asian Americans," "Socioeconomic Position and Health Behaviors

Among Californians," "Health and Work: The Status of Healthy Aging in California," "Employment Status and Health Outcomes: Extensions of the Socioeconomic Status–Health Gradient Model," "Full-Time Versus Part-Time Work: The Health of Women in California," "Employer-Based Health Insurance For Californians: Fact or Fiction?," and "Health and Employment Profiles of Ethnic Minorities."

Topics explored by research grantees in 1999 included "The Impact of Mental and Physical Health on Racial Group Differences," "The Effect of Mental and Physical Health on Labor Market Outcomes," "Health Disparities Among Californians," "Health Effects of Different Employment Statuses," "The Association of Work-Related and Personal Stressors to Perceived Stress and Depression," and "A Study of Job Lock in California."

Topics addressed by research grantees in 2000 focused on issues relevant to workers over the age of 50, and explored these issues within three broad categories: economics, demography and health, and work and retirement. Key topics included "Employment and Health Among Latino Immigrants," "Employment and Economic Well-Being of Californians Over Age 50," "The Demographic Structure of California," "Work and Family Issues Among Older Californians," "Health Status and Health Dynamics Among U.S. and California Residents Over Age 50," "California's New Economy and Issues, Prospects, and Problems for Older Workers," "Employment-Based Health Insurance Among Californians Aged 50–64," and "The Effects of Pensions, Health, and Health Insurance on Retirement."

Key Trends Identified. Very few of the research projects conducted studies that directly informed each of the three broad trends identified by the Work and Health Advisory Panel. The March 2001 report on the future of work and health prepared by IRUS was the primary report to address these issues directly. Key findings from the IRUS report pertaining to these trends are summarized in Table 8.2.

Key Trends Identified by California Work and Health Survey. A key aim of the CWHS was to examine the relationship between work and health over time. A central finding to emerge was that poor health is both a cause and consequence of employment problems. For example, among persons who were employed in 1999, those in fair or poor health were more than twice as likely as those in better health to lose their jobs by the time they were interviewed in 2000 (17% vs. 7%). Even after taking into account characteristics known to affect employment such as age, gender, and race, persons in fair or poor health were still twice as likely to lose their jobs (15% vs. 7%). Persons who lost their jobs between 1998 and 1999 were three and a half times as likely as those who maintained employment to experience worsening health (18% vs.

TABLE 8.2

Status of Key Work and Health Trends After 5 Years

Key Trends Identified By Work and Health Advisory Panel

Getting left behind by a changing economy	*A widening of income inequality*	*Changing contract between employer and employees*
There was strong evidence in the mid-1990s that wages and incomes for low skilled residents had fallen during the preceding two decades. The strong connection between income and health status made this trend of great concern to the advisory panel.	A relatively new body of research demonstrates that wide gaps between rich and poor, even in relatively affluent societies, are associated with poor health. The evidence of the mid 1990s was that income inequality in California was large and growing.	Over the past three decades, there have been many changes in work arrangements from rapid growth in the share of women who work, to changes in access to benefit and training programs, to a wider array of work schedules and new employer-employee contractual arrangements. These changes have placed new pressures on families to combine work and family responsibilities and created new insecurities about "social contract" issues like benefits, child care, and opportunities for career advancement.

Status of Trends at the End of 5 Years (IRUS*)

During the past 5 years California experienced a period of strong economic growth, resulting in raising incomes for most residents. Falling unemployment rates have been accompanied by strong job growth. Real income has risen broadly during the past five years. There are fewer Californians "left behind by a changing economy" than there were 5 years ago.	Unemployment rates for Black and Hispanic residents reached the lowest levels on record during 2000. Real wages for the bottom 20% of California workers rose by nearly 10% for the 1996–1999 period—slightly better than the state average. The poverty rate in California fell as a result of the state's strong job and income growth. While the past 5 years have been positive in terms of unemployment and wages, many low-wage workers are still behind in terms of real wage growth for the past two decades as a whole.	There has been a reversal in some of these trends and virtually no change in some areas like health insurance coverage. With the exception of temporary help agency jobs, which have been increasing rapidly, all of the other categories of contingent or alternative work arrangements now represent a smaller share of total jobs in California than was true 5 years ago. Five years of strong economic growth did not significantly increase health insurance coverage rates.

Note: From "Five years of strong economic growth: The impact on poverty, inequality, and work arrangements in California" by S. Levy, March 2001, *The Future of Work and Health Program.* Copyright © 2001 by The California Wellness Foundation.

5%). After taking into account differences between those who did and did not lose jobs between 1998 and 1999, the former group was still over three times as likely to experience worsening health (16% vs. 5%).

These findings go on to show that persons with fair or poor health are less likely to be employed than those without significant health problems, and are much less likely to report job advancement in the past year. Working persons in fair or poor health were found more likely to report having jobs with high demands and low autonomy, to work in environments that are prone to such phenomena as high rates of crime, noise, or trash, and to have earnings insufficient to bring their households above 125% of the federal poverty level. In addition, they also stated that their families are very likely to experience significant hardships in the ensuing 2 months. Such persons were also less likely to report that they have a pension plan or health insurance benefits through their jobs.

Knowledge Gained About the Three Trends. A random sample of respondents who attended the culminating conference in March 2001 were asked to rate the effectiveness of the overall conference in increasing their understanding of each of the three trends identified by the work and health panel. Overall, four fifths of respondents felt the conference was "somewhat effective" (57.1%) or "very effective" (22.9%) in increasing their understanding of issues pertaining to getting left behind in a changing economy. Nearly three fourths said the conference was effective in increasing their understanding of the widening income inequality (42.9% said it was "somewhat effective"; 31.4% said it was "very effective"). Finally, three fourths of all respondents found the conference to be effective in increasing their understanding of the changing contract between employers and employees (47% said it was "somewhat effective"; 29.4% said it was "very effective").

Building a Network

More than a dozen meetings were held throughout the program period that brought researchers, practitioners, policymakers, and other work and health experts throughout California together for sharing knowledge and building a network of individuals. Some evidence suggests the majority of participants found these meetings useful for networking.

Overall, three panel meetings were held in the fall of 1997 to identify the key work and health trends that were facing California and to develop a framework for funding research and convening projects. Initiative grantees were brought together at the beginning and end of the program in order to introduce stakeholders to one another and to share information. With respect to meetings, nine conferences were

TABLE 8.3
Future of Work and Health Convenings, 1996-2001

Year	Panel Meetings	Conferences	Grantee Meetings
1996	—	Two Building a Workable Future conferences (San Francisco and Los Angeles)	—
1997	9/25, 10/24, 11/20	—	—
1998	—	Future of Work and Health in	
1999	—	California, 4/23 Work and Health: Demographic Diversity and the California Workforce, Inequality, and Social Policy, 5/26 CWHS Research Conference, 12/10	FWH Grantee Convening, 3/4
2000	—	Employment and Health Policies for Californians Over 50 Conference, 6/7 CCHI's Work and Health Policy Forum, Quality Jobs for Healthy Californians: Solutions for an Ethnically Diverse Workforce, 11/16 CWHS Research Conference, 12/8	FWH Summation Meeting for Grantees, 6/21
2001	—	Work and Health: New Relationships in a Changing Economy—California and International Perspectives, 15/16	—

organized throughout the program period. Each served to bring together researchers and others interested in work and health issues to share findings and lessons learned. See Table 8.3.

Conference Utility for Building Contacts and Networking.   The evaluation team conducted follow-up interviews with a random sample of individuals who attended the FWH culminating conference in March 2001 to explore how participation in FWH meetings was useful for networking with others involved in work and health issues. Respondents were asked how useful the conference was in building contacts. They were also asked to describe what was useful and whether these contacts resulted in further collaborations.

When asked how useful the culminating conference was in building contacts or networks for them or their organizations, four out of five rated the conference as "somewhat useful" or "very useful" in building contacts or networks for either themselves or their organization. Fourteen percent felt it was "not very useful" or "not at all useful." Two respondents did not recall well enough to respond (6%). When further asked if these connections had led to any collaborative work, more than one fourth (28%) indicated that these connections did result in collaborative work. The majority of the respondents (72%), however, indicated that the conference did not lead to any collaborative work.

Sixty-nine percent of the respondents provided additional comments to qualify their usefulness ratings. Although only 13% of this group identified newly established collaborative efforts as a result of their conference attendance, half of the respondents (50%) restated that the conference had not led to any new collaborative efforts. Twenty-nine percent noted that, although the conference did not lead to any new collaboration, it did maintain or strengthen the collaborative relationships that already existed. One fifth said they looked forward to possible collaboration taking place in the future. In addition, one respondent commented that even though no collaboration had taken place as a result of the conference, "receiving e-mails, newsletters, and reports had been extremely valuable." Another respondent stated, in regard to forming collaborations from the conference, that it was "not my purpose in being at the conference."

Building Knowledge

The FWH program built a unique database based on the CWHS linking work and health issues in California. In addition, the vast majority of the participants of the culminating conference who were interviewed reported that the sessions were "somewhat effective" or "very effective" in identifying linkages between work and health.

To begin examining the type of knowledge generated through the FWH program and its usefulness in helping others understand the connections between work and health, the evaluation team conducted follow-up interviews with a random sample of individuals who attended the culminating conference in March 2001. Respondents were asked several questions pertaining to (a) what key messages were heard at the conference, (b) how effective the sessions were at identifying connections between work and health, and (c) how useful they found the information presented.

Effectiveness in Identifying Work and Health Connections. A random sample of respondents who attended the culminating conference in March 2001 were asked to rate the effectiveness of each conference session in identifying connections between work and health. Overall, two fifths could not recall the sessions well enough to rate their effectiveness. Among those who could recall, over two thirds rated all sessions as "somewhat effective," or "very effective," in identifying links between work and health. More than one in ten (11%) felt the sessions were "not effective."

Sessions found most effective included (a) "The Corporate Role of Creating Healthy Work," (b) "Community Dialogues on the Environment and Economic Development," (c) "Measuring the Intersection Between Work and Health," and (d) "The Digital Divide and the Role of Technology."

Sessions found least effective included (a) the opening general session from day 1 that discussed the California Economy Train, (b) the afternoon session on day 1 covering the changing social contract between employers and employees, and (c) the opening general session from day 2 that discussed work, health, and ethics, and the impact of programs on the health of the poor in Bangladesh.

Key criticisms of sessions centered on the need for stronger facilitation, greater cohesion among presenters, greater involvement of the audience, and a deeper exploration of the topics discussed. Other criticisms indicated that the information presented was not new, or not clearly presented. Consistent with this, factors that respondents considered as contributing to effective sessions suggested that the more effective sessions were well-organized and presented, and touched on new and important issues.

Usefulness of Knowledge Gained.   To describe the usefulness of the knowledge gained from the conference sessions and workshops for their own work, four fifths of the respondents assessed the knowledge gained as "somewhat useful" (50%), or "very useful" (30.6%). Thirteen percent did not find the information useful, and 5.6% could not recall the sessions well enough to respond. When asked to clarify the way(s) in which the information was useful, respondents reported that they gained exposure to issues and others working in this area, gained new information and an increased understanding of work and health issues, felt their own work and knowledge was validated, and were inspired to increase their focus on the impact of employment conditions (see Table 8.4).

Among those who gained exposure to issues and others working in this area, many comments reflected that exposure to others in the area was useful. One respondent gained an increased sense of connection with others, and stated, "I did not feel as if I was working in isolation." With respect to gaining new information, most feedback suggested that they came away with an increased awareness and understanding of work and health issues. As noted by one respondent, "it contributed to our work by increasing our knowledge." Several suggested that the information was not new, but was useful because it validated their existing knowledge and current work in this area. One respondent expressed this as, "I have already been involved in these issues; this session simply confirmed my existing knowledge."

TABLE 8.4

Key Ways Information Has Been Useful to Work of Respondents

Types of Usefulness	% (n)
Gained exposure to issues and others working in this area	26.9% (7)
Gained new information; increased understanding	23.1% (6)
Validated existing knowledge and work being done	19.2% (5)
Increased focus on impact of employment conditions	15.4% (4)
Increased awareness of global issues	11.5% (3)
Conference needed different focus to be useful	11.5% (3)
Already knew this information	7.7% (2)
Reported on California Work and Health Survey	7.7% (2)
Not useful	7.7% (2)
Other (miscellaneous comments)	30.8% (8)
Total Number of Respondents:	*26*

Another way respondents found the conference useful was by motivating them to increase their focus on work issues. These ranged from general comments about "labor perspectives" to specific topics such as "flex workers." In addition, a few indicated that the information was useful in broadening their awareness of these issues at a global level. For these respondents, it was helpful to be made aware of others who faced similar concerns around the world.

Among the few who did not find the information presented at the conference useful, the majority attributed this to the way the conference was organized. One respondent indicated, "It was unclear who the target audience was. I would have liked to have seen more discussion between labor and environmental folks." Another stated, "The session would have to have been divided into different professions and industries for information to be more useful. Trends are different based on the different segments." Overall, comments seemed to reflect that the usefulness of topics addressed depended on the needs and perspectives of different professional and industry groups that attended.

Work and Health Connections Not Covered. When asked if there were other important connections between work and health that respondents felt were not addressed at the conference, over half (58.3%) agreed there were important work and health connections not dealt with at the conference. One third said there were not, and 8.3% could not recall. Follow-up responses revealed a variety of connections that they wished were addressed. Sample comments included:

- Links between health and poverty (not just poverty statistics).
- Health issues among elementary school children.
- Health links to jobs lacking benefits and adequate compensation.
- Environmental health risks.
- Impact of upcoming recession in California (and nation).
- Demographic disparities in health insurance coverage.
- Gender differences in work and health links.
- Occupational health and safety links to work and health.
- Issues impacting psychological health and well-being of workers.
- Links between medical and employment models.
- Health issues among the welfare population.
- Both employer and employee perspectives on work and health connections.

Although more than half felt that there were important issues not addressed, three fourths of all respondents felt the conference was "somewhat useful" or "very useful" for helping them identify directions for future research on work and health issues.

Identifying Policy

The principal policy-relevant findings identified within the FWH program were developed by the California Center for Health Improvement (CCHI). CCHI produced six policy papers providing more than 40 different policy recommendations. Topics included (a) the impact of the changing California economy on the future health of workers and their families, (b) strategies for bridging California's digital divide and improving health, (c) opportunities to improve productivity and mental health of workers, (d) the changing retirement landscape in California, (e) the role of effective job search programs in preventing the threat of job loss on health and mental health of Californians, and (f) considerations for partially paid family leave in helping Californians balance work and family.

CCHI was funded to identify and publish a series of papers exploring key work and health connections relevant to each of the WHI programs. In each of its reviews, CCHI also offered policy recommendations for addressing key issues. These are now summarized in greater detail. Most of the research that informed CCHI and other policy efforts came from the longitudinal CWHS data set. Related to this, UCSF conducted a variety of policy-relevant analyses that were used by several outside groups, including policymakers and interest and advocacy groups. Other grantees within the program also identified policy-relevant findings. In their March 2001 report, IRUS identified several policy concerns pertaining to the changing economy, reducing income inequality, and supporting positive work arrangements. Although core concerns of policymakers were identified in this report, no policy recommendations were actually offered.

California Center for Health Improvement (CCHI). The CCHI received a grant within the WHI to develop policy briefs on the connections between work and health. Three briefs utilized data from the CWHS. In addition, CCHI organized a series of meetings with legislators in Sacramento. Specifically, CCHI arranged CWHS briefings in September 1999 with California legislators and staff members to discuss implications of findings for statewide health and employment practices.

Between December 1999 and April 2001, CCHI produced six policy papers in the series, "Exploring Connections Between Work and Health." Two of the policy papers were eight pages long, and the remaining were four pages each. Topics included (a) the impact of the changing California economy on the future health of workers and their families, (b) strategies for bridging California's digital divide and improving health, (c) opportunities to improve productivity and mental health of workers, (d) the changing retirement landscape in California, (e) the role of effective job-search programs at preventing the threat of job loss on health and mental health of Californians, and (f) considerations for partially paid family leave in helping Californians balance work and family. Together, more than 40 different policy recommendations were offered in this collection of policy papers. Additional groups using policy-relevant findings from CWHS are now described.

SB910. In support of this legislation pertaining to the aging population, SB910 provided funding to the University of California to convene experts in several areas related to aging. The CWHS team was commissioned to take the lead on employment issues affecting older workers. As a result, they wrote a briefing paper for legislators and participated in legislative deliberations to improve the employability of older Californians. In addition, the CWHS data have been used by several other academics and policy advocates who were part of the SB910 team.

San Francisco Board of Supervisors' Finance Committee. Testimony was given to the San Francisco Board of Supervisors' Finance Committee on the relationship between low-wage employment to health status and to health care access.

Latino Issues Forum, San Francisco. The Latino Issues Forum in San Francisco is an advocacy and research organization whose goal is to inform the public and policymakers of policy issues related to the

Latino community. In turn, it provides information to the Latino community that can be used in advocacy campaigns. The forum received a peer-reviewed minigrant from the CWHS project and developed a profile of health and employment among Latinos in California. The forum has used the results of this project to provide direction to their advocacy activities and legislative agenda.

Employee Benefits Research Institute (EBRI) Forum on Health Insurance. The EBRI has initiated a major program to inform policy debate in Washington, DC, about the health impacts of not having insurance coverage and about the potential health benefits of extending employer-based health insurance policies. UCSF was asked to prepare a paper to address these issues and to participate in a conference designed for government officials, advocates, health-foundation officers, and others interested in health insurance policy. The paper provided evidence on the health impacts of lacking insurance and was considered to be crucial to the policy deliberations.

Working Partnerships. Working Partnerships is an organization devoted to public education and advocacy about the impact of the emerging nature of work in Silicon Valley. The CWHS provided key information for the organization in the development of its report, "Walking the Lifelong Tightrope." The results showed the extent to which short job tenures and the absence of health insurance and pension benefits have become the norm in Silicon Valley jobs. Working Partners distributed 5,000 copies of the report and used the information to launch initiatives to develop mechanisms called labor–market intermediaries to help workers improve their skills, find out about employment opportunities, and obtain benefits that were formerly provided by employers.

Usefulness of Conference in Identifying Policy Implications. Finally, feedback received from respondents during follow-up interviews revealed that more than three fourths of those interviewed by the evaluation team felt the March 2001 conference was effective in helping them identify policy-relevant implications of topics discussed.

Dissemination

The FWH program was successful in attaining broad media coverage, which included 75 print publications, 34 stories in broadcast sources, and coverage on wire services for three out of four years. In addition, more than 50 manuscripts discussing work and health related findings were prepared for professional presentations or publication. However, the evaluation team was not able to determine how informative and/or useful these efforts were to target constituents.

Media Coverage. Key findings from both the "Work and Health of Californians" survey conducted in 1996 and the longitudinal "California Work and Health Survey" (1998–2000) were broadly disseminated each year. As part of this effort, data tables and summaries of key findings were prepared and disseminated via press releases to the public on an annual basis. Overall, the program was successful in attaining broad coverage, including:

Research Reports. Approximately 10 papers utilizing CWHS data were submitted for presentation at professional conferences. Each of the conferences included relevant work or health audiences, including the American Public Health Association, American Thoracic Society, Society for Epidemiological Research, American Industrial Hygiene Association, Association for Health Services Research, and the UCB Work and Family Conference. In addition, approximately 50 papers were produced, including those prepared during the first iteration of the program. All papers utilized the statewide data collected each year through FWH on the work and health of Californians as a primary source of information. Each of the papers commissioned by the FWH program were compiled and disseminated by FWH team members at conference proceedings. In addition, many of these reports were summarized in abstracts and disseminated to those interested (see Table 8.5).

CONSIDERATIONS FOR FUTURE PROGRAMMING

This section briefly highlights key information or lessons learned that might help improve future grant making by TCWF in this area. Future efforts like the FWH program should consider creating direct links between program goals and research activities, establishing mechanisms for integrating and synthesizing findings, including the employer perspective on work and health, linking policy and dissemination with research efforts, and requiring key grantees to participate in evaluation.

Need for Direct Links Between Program Goals and Research Activities. Much effort was put into the FWH panel process to develop a program framework that included three trends that had important implications for the work and health of Californians. Unfortunately, there appeared to be a gap between many of the projects that were funded and how well they contributed to increasing our understanding of the core

TABLE 8.5
Summary of FWH Media Coverage

Coverage Type	1996	1998	1999	2000
Print Media				
Major Print Outlets	23 publications	21 publications	16 publications	15 publications
	SF Chronicle, 9/2, 9/3, 9/4	SD Union Tribune, 9/7	LA Times, 9/6	LA Times, 9/4
	Sign On San Diego, 9/2, 9/3, 9/4	SF Chronicle, 9/7	SF Chronicle, 9/6, 9/7	SF Chronicle, 9/4, 9/5
	Orange County Register, 9/2, 9/3, 9/4	Sacramento Bee, 9/7, 9/8	Sacramento Bee, 9/6, 9/7	Sacramento Bee, 9/4, 9/5
	Daily Star, AZ, 9/2	SD Union Tribune Online, 9/7	SJ Mercury, 9/6, 9/7	SJ Mercury, 9/5, 9/6
	LA Times, 9/25	SJ Mercury, 9/7 (2)	Riverside Press-Enterprise, 9/6, 9/7	The Press-Enterprise Riverside, 9/4
	Sacramento Bee, 9/2, 9/3, 9/4	SF Gate, 9/7	SD Union-Tribune, 9/6, 9/7	SD Union-Tribune, 9/4, 9/5
	SJ Mercury, 9/2, 9/3, 9/4	The Press Enterprise Online, 9/7	Santa Rosa Press Democrat, 9/6, 9/7	Stockton Record, 9/4, 9/5
	La Opinion, 9/2, 9/3, 9/4	LA Daily News, 9/7	Bureau of National Affairs, HR Newsletter, 10/8	No. County Times-San Diego, 9/9
	LA Daily News, 9/2, 9/3, 9/4	Modesto Bee Online, 9/8	Health Affairs, 11/99	Fresno Bee, 9/4
		Orange County Register Online, 9/8	Business Week, 1/8/00	La Opinion, 9/5
		SF Gate Online, 9/8		
		UCSF Daybreak, 9/8 (2)		
		Marin Independent Journal, 9/8		
		SF Examiner, 9/8		
		SJ Mercury Online, 9/8		
		La Opinion, 9/9		
		Daily Labor Report, 9/8		
Broadcast Media	4	7	15	8

TABLE 8.5 (Continued)
Summary of FWH Media Coverage

Coverage Type	1996	1998	1999	2000
Radio	National Public Radio	KCBS, SF, 9/7 KQED FM, SF, 9/7 KNX, SF, 9/7 KFWB, SF, 9/7 KLIV, SJ, 9/18	KGO, SF, 9/6 KLIV, SJ, 9/6 KCBS, SF, 9/6, 9/7 KNX, LA, 9/6 KBFK, Sacto, 9/7 KQED CA Report, 9/6 CBS National Radio News, 9/6 Voice of America, 9/9 WGN Radio, Chicago, 9/12	KCBS, SF, 9/4 & 9/5 KNX, LA, 9/4 & 9/5 KFWB, LA, 9/4
Public Affairs Shows			SF Citivision, 9/8	KQED, Oakland, 9/4 KPFA, The Living Room, Berkeley, 9/11
Television	ABC News Kyodo News Services Pacific News Services	KRON, SF, 9/6 KPIX, SF, 9/7	KTVU, SF, 9/6 KRON, SF, 9/6 KGO, 9/6, 9/7 Bay TV, 9/6 KTSF, Ch. 26, 9/6	KTVU, Oakland, 9/4 KPIX, SF, 9/5 Bay-TV, SF, 9/5
Wire Services	1	0	3	3
Wires	Associated Press, 9/2		Associated Press, 9/6, 9/7 Reuters, 9/6	Associated Press, 9/4, 9/5 Reuters, 9/5

Note. 1996 coverage was for release of the "Work and Health of Californians" survey (TITF). Coverage between 1998-2000 was for the "California Work and Health Survey" findings (UCSF).

program themes. Future efforts would benefit from establishing explicit connections between the program framework and funded projects.

Need for Mechanisms for Integrating and Synthesizing Findings. Developing a panel composed of researchers and practitioners with expertise in work and health areas proved useful in identifying and prioritizing key trends and creating a program framework. What was missing was a continuation of the panel or another mechanism for identifying, integrating, and synthesizing findings from the funded projects. More efforts to identify the most important and sound evidence linking work and health in California could have proved to be fruitful.

The Employer Perspective Was Missing. For the most part, the perspectives of the employers of California workers appear to be missing from the FWH program. Future efforts would likely benefit from identifying and clarifying the role of employers in work and health. A key gap in knowledge exists with respect to the opportunities and constraints facing employers throughout California.

Link Policy and Dissemination With Research Efforts. A key goal of FWH was to identify policies that can influence work and health trends to improve the health of Californians. Aside from the CWHS findings, there did not appear to be any identification of policy-relevant findings from other research conducted within the FWH program. Groups such as CCHI were instrumental in working with initiative stakeholders to begin identifying policy-relevant implications of research findings. The influence of the FWH program would be strengthened by developing clear policy messages, identifying target audiences, policy opportunities, and creating clear dissemination strategies.

Require Key Grantees to Participate in Evaluation. When the FWH program was revised, it omitted the continuous improvement component of the evaluation that was present in the original grant structure (and other initiative programs). Without clear expectations or role requirements to participate in the evaluation, there were higher levels of resistance to serious evaluation efforts. As a result, most of the grantees were able to avoid meetings and conversations that might have led to the collection of useful evaluation data. If FWH grantees were required to participate in the evaluation in the same way that the other WHI grantees did, it is our view that the FWH program would have had greater impact. In general, requiring formal participation in evaluation helps to reinforce collaboration and participation in evaluation efforts, which, in turn, enables grantees to benefit from continuous improvement efforts.

CONCLUSION

The application of program theory-driven evaluation science to evaluating an evolving and rather complex research program was presented in this chapter. After much discussion and conflict about the nature of the program, a program impact theory eventually developed to focus the evaluation on questions that fell within five main areas: identifying trends, building a network, building knowledge, identifying policy, and dissemination. The basis for answering the evaluation questions deemed most important by the FWH stakeholders was formed by qualitative and quantitative data. Evaluation findings and evaluative conclusions were presented to illustrate sample products from the program theory-driven evaluation of the FWH program. The strengths, challenges, and lessons learned from the program theory-driven evaluation of the FWH program are explored in detail in chapter 12.

9

Evaluation of the Entire Work and Health Initiative

The purpose of this chapter is to illustrate how program theory-driven evaluation science was used to evaluate the entire WHI. In review, the overall mission of the WHI was to improve the health of Californians by funding four programs designed to positively influence health through approaches related to employment. The specific goals of the WHI were:

1. To understand the rapidly changing nature of work and its effects on the health of Californians.
2. To increase access to high quality employment for all Californians.
3. To improve conditions of work for employed Californians.
4. To increase the availability of worksite health programs and benefits.

This chapter is organized to provide an overview of the evaluation of the entire initiative, to present cross cutting evaluation findings and conclusions, to address evaluation areas of special interest to TCWF, and to explore considerations for future initiatives and grant making.

EVALUATION OF THE ENTIRE INITIATIVE

In accordance with the evaluation mandate, the evaluation team was funded to evaluate both the individual programs within the WHI and the effectiveness of the overall initiative in achieving its goals. Toward this end, evaluation findings and conclusions for each of the overall WHI goals are presented next. In addition, the evaluation team was asked by TCWF to address lessons learned with respect to several specific topics that were of interest to TCWF's grant-making efforts. These topics included (a) grantee capacity building, (b) effectiveness of program coordination teams, (c) program sustainability within funded projects, and (d) effectiveness of initiative program management. Evaluation findings for these issues are presented here as well.

The findings for the overall initiative were based on extensive quantitative and qualitative data that were collected and analyzed for each program. These were described in some detail in the previous chapters covering each program. Additional data collection efforts and analyses centered on identifying and evaluating crosscutting goals, synergies, and lessons learned. To accomplish this, the evaluation team held many conversations and meetings with grantees, conducted numerous observations, participated in initiative-wide and program-specific meetings, conducted extensive reviews of grantee progress reports and other documents produced within the initiative, and conducted 30- to 60-minute interviews with most of the WHI grantees that encouraged them to reflect on and share their views on the strengths and weaknesses of their programs and the entire WHI.

Before we present crosscutting evaluation findings, it is important to note that the WHI grantees were not funded or required to collect data for crosscutting evaluation efforts. Although there was general interest among grantees for identifying connections between program efforts and for discovering opportunities for creating synergy toward shared goals, there was no expectation on behalf of TCWF that these programs would be held accountable for engaging in such efforts. Consistent with this, no specific evaluation questions were developed for crosscutting efforts. Based on extensive interactions with grantees, the evaluation team was successful in identifying several crosscutting goals that represented the means by which program grantees were addressing the four broad initiative goals. As described in chapter 3, these included efforts designed to build or foster understanding, demonstration, influence, establishment, and evaluation of work and health links. As part of these efforts, we identified the amount of resources that each program allocated toward these goals, and identified strengths and weaknesses in the collective efforts of initiative programs in addressing each component. Although we gained agreement in several areas for building synergy, particularly around job development and policy influence, grantees ultimately did not pursue these opportunities because they did not have the time or resources to do so. Our findings therefore reflect the culmination of program-specific efforts that contribute to the overall achievement of initiative goals.

CROSSCUTTING EVALUATION FINDINGS AND CONCLUSIONS

Understanding Links Between Work and Health

All WHI programs increased understanding of relationships between work and health. The two research programs (HIPP and FWH) were particularly focused on identifying links between work and health around issues of health care, health insurance, employment status, and conditions of work. The two demonstration programs (CIOF and WNJ) provided more indirect evidence of what works in helping to prepare individuals for employment and their implications for improving the health of Californians.

Based on the collective input and wisdom of many initiative stakeholders, the central message that was created to help explain the link between work and health was concerned with having high quality employment. More specifically, "The best strategy for improving the health of Californians is a good job." The evidence for this was based on the vast literature as well as on data collected within the initiative that demonstrates a variety of ways in which access to and conditions of work are related to health.

A key achievement was the creation of unique statewide databases covering work and health and health insurance issues in California. A significant number of publications, presentations, research efforts, meetings, and evaluations of each program made substantial progress toward deepening the understanding of the rapidly changing nature of work and its effects on the health of Californians.

Increasing High Quality Employment

A significant number of participants in the two demonstration programs, WNJ and CIOF, reported finding high quality employment or advanced educational opportunities that might lead to high quality employment in the future.

The second overarching goal of the initiative was to increase access to high quality employment for all Californians. The two demonstration programs, CIOF and WNJ, were designed to address this goal directly by increasing opportunities for employment through technology skill building and job-search assistance. The two research programs, HIPP and FWH, contributed to this goal by identifying important organizational practices (such as providing health insurance and good working conditions) that were related to improved health opportunities for Californians.

With respect to the demonstration programs, the evidence we collected demonstrated that individuals in both WNJ and CIOF found jobs.

Nearly two thirds of program completers found jobs within 6 months of completing WNJ. As one indicator of employment quality, more than half of those who found jobs were reemployed in their chosen occupation. In contrast to WNJ, CIOF program grantees did not develop systematic employment follow-up data collection efforts. Most program leaders tracked employment success stories by asking participants who used their centers to tell them if and when they got jobs. As a result, several success stories were reported at each center over the 4-year funding period.

Both the HIPP and FWH programs engaged in systematic, comprehensive, statewide data collection efforts that contributed to the understanding of important organizational practices that were related to improved health opportunities for Californians. The HIPP demonstrated that most Californians got their health insurance through their employer, and those without health insurance had poorer health than those who had health insurance. The HIPP suggested that a key strategy for improving access to health insurance and health care would consist of employers increasing their offer rates of insurance. Another key strategy entailed efforts to increase affordability for both employers and employees. The FWH demonstrated that poor health is both a cause and consequence of employment problems and identified several conditions of work that were related to poor health outcomes.

A key limitation to all these efforts was a lack of an agreed-on definition of and/or standard for identifying what constitutes high quality employment. The notion of continuous training and workforce preparation was a recurring and important theme identified by WHI stakeholders. This was also considered a key solution to ensuring high quality employment. Follow-up interviews with some of the CIOF program leaders raised the question of the appropriateness of targeting employment outcomes for youth and young adults ages 14 to 23. In contrast, there was strong agreement concerning the need for improving educational opportunities, basic skills, and literacy among this group to ensure that youth and young adults can compete in the workplace. Among WNJ program leaders, there appeared to be consensus in viewing employment success along a continuum from getting a job, a better job, and then a career.

Improving Conditions of Work

We have no evidence to support that conditions of work were improved for employed Californians. Furthermore, there appeared to be no resources or programming specifically allocated toward achieving this goal.

The third overarching goal of the initiative was to improve conditions of work for employed Californians. Within the overall design of the initiative, there were no resources or programming allocated toward improving the conditions of work in organizations. Although the FWH program was initially charged with improving organizational practices that could lead to improved health of Californians, this goal was revised by the initial program director in 1996, then later dropped when the program was reconstituted between 1997 and 1998. Both the HIPP program, and especially the FWH program, generated findings in this area that could be used to identify organizational conditions that need improvement.

The design aspects of initiative programs that had the most potential for impacting work conditions, albeit indirectly, were the dissemination and policy influence components. The volume of reports generated and disseminated, the number of meetings held and attended, and the amount of media coverage gained, illustrate the potential range of influence the initiative had for indirect influence of work conditions. It is certainly possible, for example, that the broad dissemination efforts to the general public reached organizational leaders and informed, if not influenced, their thinking about these issues. It is also possible that individuals connected with the initiative increased organizational leaders' thinking through policy and educational efforts. Unfortunately, without resources directed toward this objective and a lack of interest among program stakeholders to examine the link between these issues, it is not possible to determine whether any of these efforts influenced organization practice.

In addition to a lack of programming in this area, a key limitation was the lack of involvement of corporate players throughout the state. The ability to affect organizational practices and the conditions of work will likely be enhanced by input from and cooperation with organizational leaders.

Expand the Availability of Worksite Health Programs and Benefits

There was no evidence to demonstrate that the WHI increased the availability of worksite health programs and benefits, nor were resources or programs dedicated to this goal. However, the HIPP tracked the availability of worksite health promotion programs and made policy recommendations to suggest how this might be accomplished.

The fourth broad goal of the initiative was to expand the availability of worksite health programs and benefits. There were no projects funded to expand the availability of health programs. The HIPP, however, was charged with supporting the development of state policy to increase access to health insurance coverage that emphasized the integration of health promotion and disease prevention. Each year, the HIPP collected data from both employers and health insurers on this topic and developed policy recommendations to affect this goal. A key finding was that most employers surveyed do not offer worksite health promotion programs in California. Although many health insurers offered these programs, low rates of utilization prevent these programs from maximizing health improvement among employees.

Evaluation of Areas of Special Interest to TCWF

A summary of key lessons learned about topics of special interest to TCWF is provided in this section. Lessons learned were extracted from numerous sources of data collected over the life of the initiative, including grantee progress reports, site visit observations, quantitative and qualitative data collected by the evaluation team, and through conversations with TCWF program officers and program grantees. In addition, the evaluation team conducted 30- to 60-minute follow-up interviews with most grantees after their programs were completed to specifically discuss lessons learned. From these efforts, we present crosscutting lessons learned in regard to (a) grantee capacity building, (b) effectiveness of program coordination teams, (c) program sustainability within funded projects, and (d) effectiveness of initiative program management.

Capacity Building

The main ways in which capacity was built among grantees included (a) expanded outreach and increased services offered, (b) increased visibility and exposure for funded organizations, (c) strengthened organizational knowledge and skills, (d) increased expertise in serving target population, and (e) strengthened connectivity with others. The ways most likely to impact capacity building included effective working relationships with technical assistance providers, usefulness of program evaluations, and personnel turnover.

Capacity building meant different things to different grantees. In general, capacity building was perceived to include the multitude of ways that organizations were improved or changed that enabled them to do things they otherwise would not be able to do. For some, capacity building was strengthened from acquisition of additional resources; for many, capacity building resulted from gaining new knowledge, skills, and experience. Several crosscutting lessons learned with respect to capacity building are now presented. Due to the different foci of each program, and the unique challenges and successes addressed within each program, some lessons do not apply equally to all initiative grantees. The goal here is to illuminate key ways that capacities were strengthened throughout the initiative.

Resources Expanded Outreach and Increased Services Offered.   Most grantees understood that funding was based on the expertise and experience they possessed. In light of this, the most frequently mentioned way they felt their capacity was built was through enabling them to expand their services to reach new populations. Among program managers, this was frequently expressed as "getting a chance to apply what we do in these types of organizations." The result was to strengthen their understanding and effectiveness in working with these types of (community-based) organizations. As noted by one technical assistance provider, "We gained a greater depth of experience in working with service organizations. We developed a much deeper understanding about the effect of the program on organizations that adopt them." Comments by program grantees were centered more directly on their increased ability to serve their target population through expanded and improved programming efforts.

Increased Visibility and Exposure for Funded Organizations.   Many grantees stated that their involvement in the initiative led to increased visibility for their organization. As a result, their positions as leaders in their field were viewed as strengthened. The components that seemed to contribute most directly to increased visibility included attention generated from dissemination and policy influence efforts. In addition, exposure to others through meetings, discussions, and publications of program findings also served to increase their visibility. Some grantees noted that funding by TCWF showed other funding sources that their organizations were worthwhile investments, which was also a confidence booster for the grantees.

Strengthened Organization Knowledge and Skills.   Building new knowledge, skills, and expertise was a core way in which many grantees developed capacity. Related to this, many differentiated between building personal skills versus building organizational skills. On one hand, their personal knowledge and skills were often strengthened, enabling them to do new things or improve their programs. However, institutionalization of these skills was not automatic. Key challenges for institutionalizing knowledge and skills included (a) personnel turnover, (b) lack of documentation of lessons learned, (c) lack of staff development efforts, and (d) a lack of visibility and support from senior management.

Increased Experience and Expertise in Serving Target Populations. Funding from TCWF gave many grantees the opportunity to work with new or hard-to-reach populations, enabling them to put to the test many of their beliefs and assumptions and to learn more about how best to serve their target population.

Connectivity With Others Was Strengthened.   The very design of the initiative placed a high value on collaborative relationships. This was viewed as central to the design, implementation, management, and continuance of programs. Through numerous meetings, conference calls, and electronic exchanges, initiative grantees increased their sense of connectivity with others who were faced with similar challenges and opportunities. Additional grants to CompuMentor were helpful in providing support for Web site development for initiative programs and providing assistance with creating electronic discussion lists. Overall, an increased use of technology played a key role in enabling grantees to communicate with one another, share information, and seek help from each other. As a result, a strong network emerged among initiative grantees.

Effective Working Relationships Were Essential to Building Capacity. Most grantees noted that their technical assistance providers were terrific to work with and made real contributions to their organizational planning and improvement efforts. Not all technical assistance providers, however, were perceived as having or offering needed skills and expertise. Some grantees wanted the ability to select their own technical assistance providers from their local communities. Factors noted as important in facilitating useful relationships between technical assistance providers and grantees included being within close proximity of one another and sharing similar backgrounds and/or personalities. Establishing trust and rapport were seen as the most important factors for building effective working relationships.

Evaluation Was Instrumental in Building Capacity.   Initially, program stakeholders did not necessarily see the connection between program evaluation and the use of findings for facilitating program improvements. Many resisted the evaluation and some found the evaluation process to be burdensome. To be effective, stakeholders must understand how the evaluation fits in with their overall program goals and how they will benefit from incorporating evaluation practice into their organizations. Many grantees said they gained a new understanding of evaluation. Some reported that this was the first time that they had understood the value of evaluation, and others said this was the first time they had useful data. They also reported gaining new skills in gathering, implementing, and utilizing data for program monitoring and improvement efforts. Evaluation capacity building was especially useful in program dissemination efforts that included marketing in general, grant proposals, and policy influence.

Staff Turnover Impacted Capacity Building.   Each of the initiative programs experienced some level of turnover of key personnel. Turnover resulted in the loss of programmatic knowledge and skills that were developed in staff, and often resulted in interrupted service to target constituents. In some cases, turnover resulted in the loss of program resources, such as curricula and data collection resources. Newcomers were not often informed about the history or nature of the program they entered, and did not share the same vision or understanding of the program. In some cases, turnover resulted in lack of understanding and commitment to key program goals among new hires. Therefore, it is important to expect and plan for turnover in these types of organizations, and identify strategies for minimizing the impact of turnover on capacity building efforts.

Effectiveness of Program Coordination Teams

Within the WHI, program-coordination teams served dual roles of providing guidance and technical assistance. Program coordination teams were generally viewed as very effective by program grantees. Aspects of their role that grantees viewed as most important included (a) providing direction and focus in programming efforts, (b) providing advocacy and support to grantees, and (c) organizing and facilitating meetings. Direct technical assistance provision and role conflicts were viewed less favorably.

Providing Direction and Focus Was Important.   A key function fulfilled by program coordination teams was to keep grantees focused on their program goals. Program coordination teams also helped to facilitate grantees' sense of being connected to something bigger than individual programs and/or organizations. Because program coordination teams were usually the main contact grantees had with the overall initiative, there was often confusion about their roles in relation to TCWF program officers during the early phase of each program. This was exacerbated when program managers and TCWF program officers occasionally sent conflicting messages to them.

Advocacy Played an Important Role. Most grantees felt their program coordination teams played an important advocacy role on their behalf. This ranged from efforts to secure additional funding to providing support for developing dissemination materials and influencing policymakers and other community leaders on their behalf. As a result, most grantees commented that they felt very supported by members of their program management teams. As noted by one grantee, "They really cared about us and took the time to give us the support we needed." Similarly, another grantee recalled, "They didn't have to be that way—they could have been very procedural and task-oriented. Instead, they really seemed to care about us." Advocacy, however, sometimes got in the way of providing strong guidance and leadership. It is the evaluation team's view that some technical assistance providers advocated too vigorously and were overly concerned with maintaining supportive and friendly relationships—at the expense of acknowledging problems and working with grantees to address program weaknesses and challenges.

Organizing and Facilitating Meetings Added Value. Program coordinators were found especially effective at organizing and facilitating meetings, conference calls, and meetings among the grantees. Many grantees said they did not have the time, skills, or desire to take on this responsibility. They were very happy to have a third party fulfill this role and help facilitate decision making and collaboration across sites. Related to this, program coordination teams were found to be instrumental in pushing grantees to discuss and address important issues and program challenges. In addition, they encouraged and reinforced the importance of cross-site collaboration. Several grantees noted that they would not have felt they had something to offer other grantees if they had not been encouraged to share their practices and lessons learned. Others said they would not have been as receptive to learning from others had it not been reinforced by their program managers.

Unmet Expectations Reduced Perceived Effectiveness. Unmet expectations posed an important barrier to grantees' willingness to work with program coordination teams. This was especially a problem during the start-up phase of each grant. Particularly within the CIOF program, many grantees held expectations about what they thought their program coordination team would do for them. When these expectations were not met, the result was sometimes frustration, disappointment, and withdrawal. Several grantees perceived that they did not receive equitable levels of technical assistance support compared to their peers. A key factor that impacted these relationships was geographical distance between the program manager and the implementation site. Other factors included a lack of role clarity and a lack of expertise that was needed by grantees. For example, in terms of role expectations, many sites expected that CompuMentor would provide them with more hands-on technical assistance, and were disappointed that this did not happen. With respect to types of technical assistance needed, many CIOF grantees noted they wanted more expertise from their program coordinators in fund raising, curriculum development, employee development, and staff development.

Sustainability

Several months out from their funding period, we see evidence for many sustained practices among several WHI grantees, but little sustained programming among others. Key challenges to program sustainability include (a) unclear goals and expectations, (b) a lack of early and sustained resource procurement efforts, (c) an insufficient evidence base, (d) a lack of top management support, and (e) limited availability of core operating support from funders.

Sustainability may occur along a continuum ranging from (a) continued offering of programs as designed, (b) continued offering of certain program components, and (c) to new behaviors and skills developed through funded efforts in other organizational practices. The program that appears to have the most promise is CIOF. Most centers continue to offer some level of open access and training, although the hours of access and focus of training may have shifted to match new funding opportunities and constraints. Through additional funding from TCWF, CIOF grantees continue to function as a statewide network, with a focus on policy influence activities, joint fundraising efforts, and continued support of each other. Nine of the original CIOF grantees received funding through the Workforce Investment Act (WIA) legislation to provide ongoing multimedia training to their communities. In contrast, among the three WNJ sites, only one (Proteus, Inc.) continues to offer the WNJ program as part of its Rapid Employment Program. Finally, researchers from both FWH and HIPP programs have indicated that they continue to conduct research in the same areas that they were funded by TCWF, and that they are seeking new funding opportunities to carry on some of the same types of research.

There are numerous challenges to program sustainability. Highlighted now are several key lessons learned with respect to sustainability within the WHI.

- Unclear goals and expectations about sustainability can delay or prevent sustainability.
- Resource procurement needs to start early and build over the life of the project.
- Sustainment requires strong top management support.
- Institutionalization may be an unrealistic goal for some programs.
- Demonstrating success and value added is critical for convincing external funding sources.
- Developing relationships with corporate sponsors is time consuming and requires a liaison with a high level of business acumen.
- Many funding sources look to fund only new programs.
- It is easier to get funding to support infrastructure needs (e.g., technology, curriculum) than core operating support, technical assistance, and staff development.

Effectiveness of Initiative Program Management

Finally, one of the unique components of the evaluation was the 360-degree evaluation feedback process that was implemented to give grantees an opportunity to give feedback on the effectiveness of initiative program management. Evaluation of TCWF program officers by grantees during the first 3 years of the initiative revealed several factors that contributed toward successful management of the overall initiative. These included TCWF sensitivity and responsiveness to the concerns and needs of grantees, and demonstrated interest and involvement of program officers in grant programs. In addition, grantees viewed TCWF program officers to be very accessible, approachable, and hands-on. Program officers were also commended for being solution-focused, rather than problem-focused. Grantees also suggested several areas for improvement, including (a) providing better clarification of roles and responsibilities, (b) providing more education on how foundations operate, (c) providing clear communications about changes in foundation policies or program expectations, and (d) hiring a professional facilitator for meetings of grantees.

CONSIDERATIONS FOR FUTURE INITIATIVES AND GRANTMAKING

This section is intended to briefly highlight key information or lessons learned that might help improve the design of new initiatives in this area. Considerations for program design and planning, program improvement, program management, and program institutionalization and sustainability are offered next.

The Initiative Enhanced Visibility and Resources for Programs.   Several of the program leaders reported benefits from being part of a larger initiative. For example, a wide variety of WHI meetings enabled leaders to share experiences, expertise, and resources with each other, to expand their professional networks, and to enhance the visibility of their programs in new domains. Several regret that they did not create and capitalize on more opportunities to create synergies across the four programs.

It Was Difficult to Motivate Program Grantees to Work on Initiative Goals.   Although there appeared to be great interest in working together to develop and achieve WHI goals beyond the program level, it was difficult to motivate program grantees to follow through. We believe this was largely due to their focus on the demands of the individual programs and because they were not funded and lacked incentives to work on the broader WHI goals. Furthermore, this was clearly deemphasized by TCWF after the first 2 years of the initiative's start.

Staff Turnover Was Frequent and Reduced Productivity.   Within each program there was staff turnover. Turnover in demonstration programs was especially frequent and occurred at all levels, from executive directors, program directors, to staff and volunteers. A major negative consequence of turnover in the WHI was that new staff typically required extensive training about the various aspects of the WHI, as well as their specific job responsibilities, to be optimally effective. It was very difficult to meet this continuous training need. Therefore, it is important to expect and plan for turnover in future initiatives in this area, and to identify strategies for smoothing over staffing transitions.

Technical Assistance Was Key in Addressing Unfamiliar Territory. Technical assistance was instrumental in helping grantees develop and implement programming in areas where they had little prior experience. Key areas in which additional technical assistance or resources for technical assistance were needed included (a) creating dissemination and communication strategies, (b) developing curricula, (c) creating policy agendas and policy influence strategies, (d) employment development, and (e) resource procurement. Unfortunately, existing technical assistance providers did not always have requisite skills in these areas (e.g., curriculum development) or were located at great distances from some grantees. As a result, there was a strong preference among grantees to use a discretionary pool of resources for meeting local technical assistance needs. However, many grantees noted they needed assistance in locating and evaluating these resource providers.

Relationship Building Was Critical to Program Success in Core Areas. Complex problems often require complex solutions that are best addressed through a partnership of resource providers. Establishing and maintaining collaborative relationships, however, can be challenging even under the best circumstances. Ongoing relationship building was critical for establishing buy-in and securing ownership with other partners. Establishing a shared understanding of and agreement around roles and responsibilities with organizational partners was an important lesson learned in the initiative.

Quality Was More Important Than Quantity of Direct Service. There exists an inherent tension between the quantity and quality of service delivered. Several grantees felt there was undue pressure to reach service goals at the expense of developing and offering high quality services to program participants. Related to this, it is important to determine how to improve direct service, and to make sure that resources are not being wasted on ineffective or potentially harmful services.

Dynamic Programs Required Dynamic Program Evaluation. Like organizations, programs develop through a progression of life stages. Complex programs go through many changes from program design to institutionalization, and evaluation efforts should be adapted to address changing information needs over time. To be responsive to these changes, dynamic evaluation requires that evaluators work closely with program stakeholders over time, are flexible in their approach to addressing changing needs and changing goals, and are able to shift evaluation resources to match new priorities. Clear communication about the role of evaluation at each phase of the evaluation is critical to fostering ongoing trust and rapport.

Institutionalization and Sustainability Goals Were Not Achieved. There was strong interest in sustaining programs beyond TCWF funding. Key challenges consisted of (a) unclear institutionalization goals, (b) a lack of guiding mechanisms for deciding which elements were worthy of institutionalization, (c) a lack of resources for planning and strengthening sustainability efforts, and (d) a lack of expertise in procuring resources and responding to Requests for Proposals (RFPs). Institutionalization and sustainability were least effective when efforts to achieve these goals were started toward the middle or the end of program periods, rather than at the beginning. Continued, sustained efforts to build top management buy-in and finding resources seem essential for ensuring programs continue beyond initial program periods.

Timing Was Very Important.   The state of the economy in California and the challenges and opportunities facing each program area looked very different 5 years ago. Historical trends and variable changes over time affected both the visibility and success of initiative programs. When CIOF was instituted, for example, the digital divide was not present in everyday vernacular. Whereas very few community technology centers existed in the mid-1990s, there was an explosion of these types of organizations during the last 5 years. CIOF policy influence efforts were being developed during a critical window of opportunity that simply did not exist in other programs. With respect to WNJ, the economy improved dramatically during the past 5 years, making it difficult for grantees to meet their target service goals (there were not as many unemployed!). Future programming should consider how macro-changes in the environment and economy will influence the success of programming efforts.

CONCLUSION

The main purpose of this chapter was to provide a summary of key evaluation findings and conclusions for the TCWF's WHI. These evaluation findings and conclusions were based on analyses of extensive quantitative and qualitative databases designed and managed by the evaluation team. Many of the findings and issues presented throughout chapters 4 through 9 are described in much more detail in one or more of the 200 evaluation reports that were written by the evaluation team and provided to WHI grantees and/or TCWF throughout the life the WHI. These chapters reflect the evaluation team's candid summative evaluation of the WHI from an external evaluation perspective. Despite the shortcomings and challenges noted throughout these evaluations, collectively, the findings of the program theory-driven evaluations of TCWF's WHI suggest that this initiative improved the lives of many California workers and their families. Evaluation evidence also suggested that this initial impact could multiply and expand over time, and that the WHI could ultimately be viewed as a successful, groundbreaking effort in the changing domain of work and health.

In chapter 10, we shift our focus away from the WHI and on to the Bunche–Da Vinci Learning Partnership Academy Case. Nicole Eisenberg, Lynn Winters, and Marvin Alkin present a case scenario in which evaluation is required. In chapter 11, I illustrate in some detail how program theory-driven evaluation science could be used to provide practical and cost-effective external evaluation services to help Bunche–Da Vinci stakeholders. This case presentation and evaluation proposal are part of a larger project to compare and examine how different theoretical evaluation approaches get moving in evaluation practice (Alkin & Christie, 2005). These final two chapters of part II are intended to illustrate how program theory-driven evaluation may be presented in a proposal format to evaluate new cases.

10

The Case: Bunche–Da Vinci Learning Partnership Academy[1]

Nicole Eisenberg
University of California, Los Angeles

Lynn Winters
Long Beach Unified School District, California

Marvin C. Alkin
University of California, Los Angeles

Bunche–Da Vinci Learning Partnership Academy is an elementary school located between an urban port city and a historically blue-collar suburb. The dignified and well-maintained school structure is a much-sought-after refuge for students, both African American and Latino, from the neighboring smaller suburb, well known for poverty, crime, racial tension, and low performing schools. The school is a beacon of hope and stability in an industrial area of the large city and is in sharp contrast to the nearby schools of the smaller, more notorious,

---

[1]This chapter is reprinted from "The Case: Bunche–Da Vinci Learning Partnership Academy" by N. Eisenberg, L. Winters, and M. C. Alkin, 2005. In M. C. Alkin and C. A. Christie (Eds.), *Theorists' Models in Action: New Directions for Evaluation* (Vol. 106, pp. 5–13). Copyright © 2005 by John Wiley & Sons, Inc. Reprinted with permission of John Wiley & Sons, Inc. and the authors.

adjacent suburb. Bunche is a "lighthouse" created by a unique partnership between the district and a nonprofit educational company specializing in innovative school interventions for low-performing students. The district recognized early on that to combat the problems that Bunche's students faced, innovative solutions were needed. The school is characterized by students with a high rate of transience, illegal enrollments from the adjacent district, and high numbers of non-English-speaking students, a young and inexperienced staff with high turnover, and geographical isolation from the rest of the district. When approached by Da Vinci Learning Corporation, the district chose Bunche Academy to enter into a unique arrangement with the company to create a partnership that combined elements of a charter school while still remaining a regular part of the district.

THE COMMUNITY

Bunche–Da Vinci is located in an industrial zone on the northwestern edge of a city of half a million. Students travel several miles and must cross busy boulevards to reach the site; thus, most arrive on district school buses. The homes in the community are mostly small ranch houses from the 1940s and boxy modern 1950s apartment buildings. Many of the dwellings house multiple families or families with several children. There are no supermarkets, department stores, or large shopping malls in the community. Most shopping is done in small, expensive mom-and-pop grocery stores, located in corner strip malls housing the predictable fast-food chains, liquor stores, and check-cashing services found in poor neighborhoods.

THE SCHOOL

Bunche–Da Vinci Learning Partnership Academy remains part of the large urban school district but is also a partnership with the Da Vinci Learning Corporation. Da Vinci is a "full-service school reform operation" that has been partnering with Bunche for the past 3 years. Schools owned or operated by Da Vinci must adhere to its curriculum, schedules, and class size requirements. Bunche is the only school working with Da Vinci that retains its autonomy from the corporation and is wholly run by a school district. Teachers work 205 days (as opposed to the regular teacher year of 185), and the school year is 200 days (as opposed to 180). Students attend an extended day, 6 to 8 hours in length. Classrooms for kindergarten through Grade 3 are staffed at a ratio of 20:1; Grades 4 and 5 are staffed at 35:1, with a 20:1 ratio during the extended reading period.

Students are grouped and placed by ability, regardless of grade level. This means that during reading and math, students are grouped by the grade level at which they are functioning (e.g., a fifth grader reading at the first-grade level goes to the Grade 1 reading group). Every student receives a full curriculum: reading, math, language arts, science, social studies, music, art, physical education, modern foreign languages, and values education. The elective program is handled in two ways. Reading teachers who are specialists also teach music, art, physical education, or modern languages. Homeroom teachers teach math, language arts, social studies, science, and values education. Students receive specialized instruction one to two periods a day. K–2 teachers do planning outside their school day. Teachers in Grades 3 to 5 have a planning period built into the school day, which runs from 8:30 to 4:30.

Da Vinci Learning focuses heavily on providing technology-rich environments for students and teachers. All teachers receive personal laptops and personal digital assistants (PDAs) while working at the school. All grade reporting and classroom bookkeeping are done online and hosted at Da Vinci. Students have multiple opportunities to work with computers and take computer-assisted tests beginning in Grade 2. The school's mission statement is: "Excellence for All; Excuses for None."

The combination of state and federal funding provides Bunche with an annual operating budget in excess of $8 million. The school, with 1,150 students, receives a $5,200 per pupil allocation yearly from the state. These state revenues are used mainly to cover salaries. The remaining operating funds come from indirect services provided by the district or from categorical funds. Specifically, Title I funding adds an additional $2 million. Education Foundation gifts, Title III, and grant funds make up about 10% of the school's budget. Da Vinci has full control of the funds as part of its partnership agreement and uses them to staff lower class sizes, fund the extended school day and year, provide technology, and pay teachers.

Enrollment at Bunche has been steadily rising over the past few years due to an influx of Latino families into this historically African American neighborhood. Many of the new Latino families are fleeing the adjacent suburb in search of safer neighborhoods and better schools. Prior to the partnership, Bunche enrolled about 1,050 students. With the advent of the partnership, enrollment dipped to 900, but in the second year, which coincided with an influx of Latino immigrants, it surged to 1,150 and is steadily growing by about 100 students a year. Bunche has always served low-income families, but the demographics have shifted dramatically over the past ten years (see Table 10.1).

TABLE 10.1

Bunche Enrollment Trends, 1995–2004: Percentage Distribution by Ethnicity

	Hispanic	African American	Asian, Filipino, Pacific Islander	White	American Indian
1995	54.2	25.2	16.3	3.5	.8
2004	78.7	17.2	3.3	.7	.1

Bunche has a large portion of students who are not fluent English speakers. Almost 60% of the school population is classified as English language learners (ELLs), with 97% of those being Spanish speakers. A little more than half of these ELL students are mainstreamed—that is, they are part of the regular classes—with extra support provided. The remaining students are enrolled in Structured English Immersion for a portion of the day and then mainstreamed with support for the rest of the day. Aside from the Structured English Immersion program, there are no other "pull-outs." That is, Title I and other funded programs are school-wide programs, where students remain in class and are not pulled out for special instruction.

Student test scores at Bunche have been among the lowest in the district for many years. Teachers note that issues of how to teach ELLs dominate faculty lounge chat and faculty meetings. Secondary concerns about the neighborhood and classroom tension between African American and Latino students, even at this young age, also consume time and attention and remain the focus of the values education program. However, students apparently are delighted to come to and remain in school, a place where they feel safe, have access to computers, and participate in organized after-school arts activities.

A NEW PRINCIPAL

Bunche was converted from a somewhat bland elementary school in a remote corner of the district to an innovative experiment in public-corporate partnership. With the advent of the partnership, the district assigned a new principal to Bunche–Da Vinci, Mary García, in hopes of signaling to the community that this was a fresh start. García was a first-year principal but hardly an inexperienced administrator. In keeping with past practice, the district assigned staff to the principalship only after they had experience at both inner-city and suburban school sites, as well as the central office. García had spent 7 years teaching at a nearby K–8 school in both the upper elementary and middle school grades. She had spent 3 years in district office assignments in both the Special Projects (Title I, Title III) and Curriculum offices, where she worked as a literacy coach. García is cheerful by nature, but beneath her ready smiles and warm manner lies fierce determination to level the playing field for her poorest and most struggling students. She has strong opinions regarding issues of equity and quality instruction and can be quite tough. In addition, she has always been known to focus on what is best for children over what is best for teachers when these values are in conflict.

Because she had spent many years working in the neighboring schools and the central office, she was well aware of the conditions and challenges facing her in her first year as principal. As is often the case in elementary schools, the staff of 72 teachers and specialists was predominantly young, and the majority had young children at home. Each year brought at least two engagements, two marriages, and three pregnancies. In terms of ethnicity, almost half the teachers and support staff were Hispanic, slightly fewer were White, and six were African American. Over half of the teachers were bilingual. Though not staffed with all beginning teachers, the teaching staff was nonetheless young. Teachers had an average age of 30 and approximately 4 years of experience.

García found that parents and students were enthusiastic about the school and supportive of the programs. Teachers, in contrast, complained of feeling "burned out" and "jerked around" by the long school day, extended year, and "imported" curriculum. The teachers' union had met with the principal at least once a year at the request of fourth-grade teachers who felt the schedule was "grueling" and the Da Vinci curriculum "constricting." Both sides were watching test scores as a way to vindicate their positions.

Staff felt they had no extra time to attend PTA meetings at night (and the neighborhood was not an inviting place to be after dark). The PTA, in fact, barely existed. A handful of parents, often accompanied by small children, met monthly to plan refreshments for the annual Teachers Day and Back-to-School Night and to schedule the monthly Parent Education meetings with the district parent education coordinator. Although the majority of parents spoke Spanish, meetings were conducted in English, and no translator was provided.

García's first action at Bunche was to make it clear in faculty meetings that a new spirit was going to be injected into the school and the new direction would be a product of what she called "joint decision making." Her belief was that to a large extent, the students' success in school would depend on the confidence and support of the teachers—the "self-fulfilling prophesy" approach. She wanted to make this prophesy positive: "We have the special skills to teach these children, and only we can turn them around." García stressed the attitude that "all roads lead to Rome" and that even with an "external" curriculum and a multitude of subjects that didn't "count" for No Child Left Behind (NCLB) accountability, Bunche–Da Vinci students could succeed.

During the partnership's first year, the new principal and her staff spent numerous sessions reviewing school data on state and district assessments and discussing possible strategies. They agreed that the school's top priority should be reading; that was clearly where students were having difficulty. Historically, this school's reading test scores had been below the state average and the proportion of students in the bottom ranges was higher than in most other schools in the district.

Early data suggested that some of their efforts paid off. During the first 2 years of the partnership, student scores on state tests showed improvement, and they were able to meet their Adequate Yearly Progress targets, as set by CLB. However, García worried because she felt this was merely surface improvement. It was true that there was some progress from year 1 to year 2. However, after the second year of the partnership, only 29% of the students scored at a level considered "proficient" in English-Language Arts, which was below the district and state average. In addition, García worried about the performance of specific subgroups of students. For example, when examining ethnic groups, she noted that only 28% of the Hispanics and 25% of the African American students were proficient. Equally disturbing was the fact that fully one third of the students were scoring in the bottom range of the state tests.

CHALLENGES AND UNFINISHED BUSINESS

Despite the initial small gains in state test scores, García felt that many problems remained at the school. The hope was that the school was on the right track. But then she received new results from the latest state test scores and found that despite the initial improvement, the partnership's third-year scores had gone down. (Table 10.2 shows the percentage of students for each subgroup scoring at the proficient level in the English-Language Arts state standards tests for the 3 years of the partnership.) The data indicate that English language scores dropped to a level below first-year scores, completely nullifying the small spark of optimism that García had felt. All groups had shown a drop in scores, with perhaps the greatest decline for African American students.

Moreover, these decreasing state test scores contrasted sharply with students' performance on Da Vinci's own testing system. Da Vinci Learning Corporation tested students on a regular basis, and on Da Vinci's standardized tests, students had improved considerably. The

TABLE 10.2

Bunche–Da Vinci Partnership Test History: Percentage Proficient on State
Standards Tests for English-Language Arts

	Year 1	Year 2	Year 3
All	28	29	24
Hispanic	27	28	25
African American	23	25	20
White	75	72	65
Asian	80	85	78
ELL	22	21	20
Low socioeconomic status	30	29	27
Special education	10	8	7

percentage of students reading at grade level had doubled over the past 3 years. Da Vinci staff from headquarters, who were sent to monitor the curriculum, indicated that they had heard students say that they were "finally able to read" and found that even quite mature Grade 5 students who were working in Grade 1 materials were enthusiastic learners. García wondered why this mismatch had occurred.

Her first thought was that perhaps it was an issue of misalignment. She knew that the school's staff often had difficulty aligning the Da Vinci reading curriculum with the state standards. The Da Vinci program emphasized sustained silent reading, literature, invented spelling, and group work. This approach diverged considerably from the state-adopted program that emphasized whole-class direct instruction, phonics, spelling lessons, and continuous assessment and reteaching.

But was it just the alignment, or was it more than that? García knew that parents and students were, overall, happy with the school. But she did not feel the same level of support from teachers. Like many other schools in low-income neighborhoods, the staff turnover was high. But these teachers were especially resentful of teachers in other district schools, even their year-round neighbors. They rarely stayed once they had opportunities to teach in a traditional calendar school. The long hours, required in-service, shorter vacation time, and additional teaching assignments were wearing teachers down despite the additional pay. Fully 25% of the teachers had emergency credentials, and another 30% were teaching outside their subject areas due to the requirements that the school offer foreign language, values education, and technology classes. With the inception of NCLB, these teachers were required to be fully certified in their subjects by taking courses or tests. The result was that fully 60% of the faculty had outside commitments after school and on weekends in order to meet these requirements. Among the faculty, 65% had been teaching fewer than 3 years. They were enthusiastic, idealistic, and committed to the proposition that all students can learn, but few had enough experience to implement a rigorous or even accelerated curriculum for students who were underprepared or were English learners. They were literally learning how to "fly the plane while building it," which created a climate of alternating high hopes and frustration depending on the day. García had a tightrope to walk between offering encouragement and raising the bar. There were very real demands on the teachers in terms of learning a new curriculum, trying to make the curriculum align to the state testing system, learning to teach, taking required courses, attending to their own small children, and knowing that they were the only school in the district with such a workload (despite being compensated for it).

García was also concerned about teachers' time on task. For one thing, she felt too much of the extended day was devoted to computer-assisted instruction. A chief goal of this instruction was to individualize instruction better, but teachers had difficulty using the cumbersome and overwhelming "diagnostic" data, which rarely mapped onto strategies and skills used by the reading program or the state standards. In the face of conflicting information about students, teachers used their own judgment and impressions to regroup students in reading and math. But the computer-based activities were only part of it. García had a hunch that teachers were not spending enough time on reteaching and accelerating the learning of the lowest performing students.

Something else was worrying García. She noted that during the partnership's second year, there were more changes in the student population than she had anticipated. This third year, she was again amazed by the number of new students and by the large bulk of students who left. Was this having any effects or negative influences on the classroom?

García knew that for the most part, parents were satisfied with the school. However, there were a few who questioned the amount of time that children spent "on the computer." Recently she had received a surprise visit from a group of six Title I parents. They had arrived in her office one morning to complain about the situation. Their children were among the lowest achieving students and were having difficulty learning to read. These parents agreed that the technology was good, but felt that it should not be "taking time away" from the "more necessary reading lessons."

García empathized with those parents. The plight of the below-grade-level student had become the principal's top concern. But she was not sure whether the parents' attributions were correct. Faculty struggled to serve these students but often could not see ways in which they could change their instruction. Not all students had made the reading gains that most of the Filipino, White, and native-born Hispanic students had, even though attendance was relatively high for all sub-groups of students. Few teachers discerned this until García pointed it out to them. Even then, faculty tended to dismiss out of hand their responsibility for the differences in reading performance. When review-ing annual test results for subgroups of students, teachers could often be heard saying, "Oh, he's never here. His family takes him to Mexico for weeks at a time." Or, "She was new to the district. Entered way behind. Of course, she scores low." And the ubiquitous, "You should see their family situation. It's a miracle that child comes regularly."

The principal felt that the school had just scratched the surface in addressing the literacy problem. Test results showed that students still struggled with spelling and editing skills, composition, and reading comprehension of nonfiction text, not to mention that mathematics had been all but ignored during the year's work on schoolwide literacy. While the Da Vinci learning curriculum was moderately effective in developing conceptual understanding, students did poorly on the state tests due to lack of fluency with number facts and procedures.

García felt that she and her staff could not tackle the school's diffi-culties alone. There were aspects about the partnership that she ques-tioned, but she lacked the skills to assess whether she was right. She felt that the school and the partnership needed to undergo some changes. She needed to think about a course of action and then meet to discuss them with the district superintendent.

After careful thought, she contacted the district superintendent, Douglas Chase. Chase was a home-grown superintendent. He was born in the community and had attended elementary, intermediate, and high school in the community. He had received a PhD in policy from a nation-ally acclaimed research university, moved across the country to teach in higher education in a major urban center, and had an extensive national collegial network encompassing foundations and government officials. He was a visionary with keen political instincts and a lateral thinker who prized diversity in thought and style. He was a passionate advocate for poor children and not content to wait generations for results. One of his favorite aphorisms was, "It's five minutes to midnight." Chase's most prominent characteristic was his novel view on age-old problems. His staff knew that whenever they asked a question, they would not be able to predict the response; it would represent a vision, a stance, a viewpoint, or a question they had not thought about before.

Chase had been at the district for many years and had witnessed Bunche–Da Vinci's changes over time. He had initially advocated in favor of the partnership, realizing that the district alone could not do much more to help schools such as Bunche. However, over time, he too had developed questions about the partnership. He had overheard some of his staff talk about the problems at Bunche–Da Vinci and was wondering whether it might be best to end the contract. With his down-to-business style, he focused on test results, which had declined in the past year. That was what he really cared about. He wondered whether the initial increase in test scores (during the partnership's first 2 years) was a result of the actual partnership and Da Vinci's "research-based curriculum," or of the district provided in-service and staff development.

García and Chase met a few times to discuss what they could do. Initially Chase wondered if the innovative partnership was part of the problem. García knew it was too early to know and asked to work with Da Vinci in a more collaborative (and less "obedient") fashion. Chase also wondered if the district needed to provide more and different support to the novel enterprise. He knew that staff—by their very attitudes—could go through the motions yet undermine the most promising of innovations. García, a can-do and well-regarded professional, felt that installing systems in place—for teacher collaboration, monitoring curriculum and instruction, student interventions, and even communication with Da Vinci—would go a long way toward righting the ship. Chase, never a micromanager, was more than willing to allow her to work the partnership her way for the coming year (or 2 at most).

Chase was open to giving García time to resolve Bunche's problems, but also wanted more data. He wanted to know why there was a disparity in Da Vinci and state test scores. He wanted to know if the Da Vinci curriculum was manageable. He had other questions as well, but he also recognized that he wanted to give García the flexibility to make changes. He wanted to support her in her efforts, and to do so she would need evaluation help, not only to monitor the impact of her changes, but also as a reliable source of evaluative data on the impact of the program or parts of it.

Chase knew that the district evaluation unit was very small and was overworked on Title and other special programs. Thus, he decided that an outside evaluation might be appropriate. He suggested that García call DGHK Evaluation Associates to ask for an evaluation and recommendations for school improvement. Chase indicated that he would find funds from one of his foundation sources for an external evaluation. García was all too eager to comply with the request for an evaluation; she hoped that it would help her in making improvements to the program.

Nicole Eisenberg is a doctoral student in the Social Research Methodology Division in the Graduate School of Education and Information Studies at the University of California, Los Angeles. Lynn Winters is the assistant superintendent of research, planning, and evaluation in the Long Beach Unified School District in California. Marvin C. Alkin is an emeritus professor in the Social Research Methodology Division in the Graduate School of Education and Information Studies at the University of California, Los Angeles.

11

Using Program Theory-Driven Evaluation Science to Crack the Da Vinci Code[1]

Program theory-driven evaluation science uses substantive knowledge, as opposed to method proclivities, to guide program evaluations (Donaldson & Lipsey, 2006). It aspires to update, clarify, simplify, and make more accessible the evolving theory of evaluation practice commonly referred to as theory-driven or theory-based evaluation (Chen, 1990, 2004, 2005; Donaldson, 2003; Rossi, 2004; Rossi et al., 2004; Weiss, 1998, 2004a, 2004b).

The purpose of this chapter is to describe in some detail how I would respond to the call from Mary García, principal of the Bunche–Da Vinci Learning Partnership Academy, asking my organization, DGHK Evaluation Associates, for a proposal to provide "an evaluation and recommendations for school improvement." Based on specific instructions from the editors, I have attempted to provide a realistic account of the actions I would take to provide evaluation services using the program theory-driven evaluation science approach. Although I have avoided the temptation of simply explicating the principles and procedures for conducting program theory-driven evaluation science again (Donaldson, 2003; Donaldson & Gooler, 2003; Donaldson & Lipsey, 2006; Fitzpatrick, 2002), I do provide a limited amount of background rationale in key sections to help readers better understand my proposed actions.

---

[1]This chapter is reprinted from "Using Program Theory-Driven Evaluation Science to Crack the Da Vinci Code" by S. I. Donaldson, 2005. In M. C. Alkin and C. A. Christie (Eds.), *Theorists' Models in Action: New Directions for Evaluation* (Vol. 106, pp. 65–84). Copyright © 2005 by John Wiley & Sons, Inc. Reprinted with permission of John Wiley & Sons, Inc. and the authors.

This exercise was a stimulating and useful way to think about how I actually work and make decisions in practice. It was obviously not as interactive and dynamic of an experience as working with real evaluation clients and stakeholders. For example, conversations with stakeholders, observations, and other forms of data often uncover assumptions, contingencies, and constraints that are used to make decisions about evaluation designs and procedures. It was necessary at times to make assumptions based on the best information I could glean from the case description and my imagination or best guesses about the players and context. The major assumptions made to allow me to illustrate likely scenarios are highlighted throughout my evaluation plan. My goal was to be as authentic and realistic as possible about proposing a plan to evaluate this complex program within the confines of everyday, real-world evaluation practice.

CRACKING THE "DA VINCI CODE"

It is important to recognize that not all evaluation assignments are created equal. Program theory-driven evaluation science, and even external evaluation more generally, may not be appropriate or the best approach for dealing with some requests for evaluation. The Bunche–Da Vinci case, as presented, suggested that one or more of a highly complex set of potentially interactive factors might account for the problems it faced or possible ultimate outcome of concern: declining student performance. Principal Mary García appears exasperated, and district superintendent Douglas Chase at a loss for how to deal with the long list of seemingly insurmountable challenges for the Bunche–Da Vinci Learning Partnership Academy. Why has such a good idea gone bad? Why is performance declining? Could it be due to:

- A changing population?
- Social groupings of students?
- Student attendance problems?
- The curriculum?
- The innovative technology?
- Language barriers?
- Culturally insensitive curriculum and instruction?
- Staff turnover?
- The teachers' performance?
- Parenting practices?
- Leadership problems?
- Organizational problems?

And the list of questions could go on and on. How do we crack this "code of silence" or solve this complex mystery? "We've got it," says García and Chase. "Let's just turn to the Yellow Pages and call our local complex problem solvers: DGHK Evaluation Associates."

It appears to me on the surface that DGHK Evaluation Associates is being called in to help "solve" some seemingly complex and multi-dimensional instructional, social, personnel, and possibly leadership and organizational problems. What I can surmise from this case description, among other characteristics, is:

- There appears to be many factors and levels of analysis to consider.
- Everyone is a suspect at this point (including García and Chase).

Some of the stakeholders in this case may have different understandings, views, and expectations about evaluation, and some may be very apprehensive or concerned about the powerful school administrators calling in outsiders to evaluate program and stakeholder performance.

The conditions just listed can be a recipe for external evaluation disaster, particularly if this case is not managed carefully and effectively. As a professional external evaluator, I do not have the magic tricks in my bag that would make me feel confident about guaranteeing Bunche–Da Vinci that I could solve this mystery swiftly and convincingly. However, I would be willing to propose a process and plan that I believe would stand a reasonable chance of yielding information and insights that could help them improve the way they educate their students. So how would I use and adapt program theory-driven evaluation science to work on this caper?

NEGOTIATING A REALISTIC AND FAIR CONTRACT

Background Rationale

In my opinion, one of the key lessons from the history of evaluation practice is that program evaluations rarely satisfy all stakeholders' desires and aspirations. Unrealistic or poorly managed stakeholder expectations about the nature, benefits, costs, and risks of evaluation can quickly lead to undesirable conflicts and disputes, lack of evaluation use, and great dissatisfaction with evaluation teams and evaluations (see Donaldson, 2001b; Donaldson et al., 2002). Therefore, my number one concern at this initial entry point was to develop realistic expectations and a contract that was reasonable and fair to both the stakeholders and the evaluation team.

The Bunche–Da Vinci Learning Partnership is a well-established, ongoing partnership program, with a relatively long (more than 3 years) and complex history. Therefore, I made the following two assumptions prior to my first meeting with García and Chase:

- *Assumption*: There are serious evaluation design and data collection constraints. This is a very different situation than the ideal evaluation textbook case where the evaluation team is involved from the inception of the program and is commissioned to conduct a needs assessment, help with program design and implementation, and to design the most rigorous outcome and efficiency evaluations possible.
- *Assumption*: Money is an object. Based on the case description, I assumed that García and Chase desired the most cost-effective evaluation possible. That is, even if they do have access to substantial resources, they would prefer to save as much of those as possible for other needs, such as providing more educational services.

It is important to note here that I would approach aspects of this evaluation very differently if money was no object (or the evaluation budget was specified), and there were fewer design or data collection constraints.

Meeting 1.   My first meeting with García and Chase was a success. I began the meeting by asking each of them to elaborate on their views about the nature of the program and its successes and challenges. Although they had different views and perceptions at times, they seemed to genuinely appreciate that I was interested in their program and daily concerns. They also said they were relieved that I began our relationship by listening and learning, and not by lecturing them about my credentials, evaluation methods, measurement, and statistics, like some of the other evaluators with whom they have worked.

After García and Chase felt that they had provided me with what they wanted me to know about the partnership, I then asked them to share what they hoped to gain by hiring an external evaluation team. In short, they wanted us to tell them why their state scores had declined and how to reverse this personally embarrassing and socially devastating trend. It was at that point that I began to describe how DGHK Evaluation Associates could provide evaluation services that might shed light on ways to improve how they were currently educating their students.

In an effort to be clear and concise, I started by describing in common language a simple three-step process that the DGHK Evaluation Team would follow:

1. We would engage relevant stakeholders in discussions to develop a common understanding of how the partnership is expected to enhance student learning and achievement. This is Step 1: Developing program theory. (Note that I rarely use the term *program theory* with stakeholders because it is often confusing and sometimes perceived as highbrow and anxiety provoking.)
2. Once we have a common understanding or understandings (multiple program theories), I explain that we would engage relevant stakeholders in discussion about potential evaluation questions. This is Step 2: Formulating and prioritizing evaluation questions.
3. Once stakeholders have identified the most important questions to answer, we then help them design and conduct the most rigorous empirical evaluation possible within practical and resource constraints. Relevant stakeholders are also engaged at this step to discuss and determine the types of evidence needed to accurately answer the key questions. This is Step 3: Answering evaluation questions.

In general, I pledged that our team would strive to be as accurate and useful as possible, as well as participatory, inclusive, and empowering as the context will allow. That is, sometimes stakeholders may choose not to be included, participate, or use the evaluation to foster program improvement and self-determination. At other times, resource and practical constraints limit the degree to which these goals can be reached in an evaluation. García and Chase seemed to like the general approach, but after they thought about it some more, they began to ask questions. They seemed particularly surprised by (and possibly concerned about) the openness of the approach and the willingness of the evaluation team to allow diverse stakeholder voices to influence decisions about the evaluation. They asked me if this was my unique approach to evaluation, or if it was commonly accepted practice these days. I acknowledged that there are a variety of views and evaluation approaches (Alkin & Christie, 2004; Donaldson & Scriven, 2003b), but pointed out that the most widely used textbooks in the field are now based on or give significant attention to this approach (examples are Rossi et al., 2004; Weiss, 1998). Furthermore, I revealed that many federal, state, and local organizations and agencies now use similar evaluation processes and procedures. They seemed relieved that I had not just cooked up my approach in isolation as a fancy way to share my opinions and render judgments. However, they did proceed to push me to give a specific

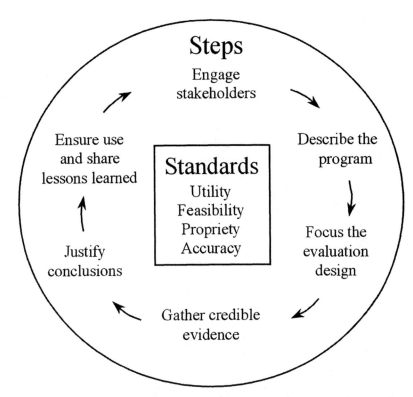

Figure 11.1.   CDC Six-Step Evaluation Framework. From *Centers for Disease Control and Prevention. Framework for program evaluation in public health.* MMWR 1999; (No. R–11). Available from http://www.cdc.gov/eval/framework#htm

example of one of these organizations or agencies. So I briefly described the Centers for Disease Control's Program Evaluation Framework (Centers for Disease Control [CDC], 1999).

The CDC Evaluation Framework is not only conceptually well developed and instructive for evaluation practitioners, but it also has been widely adopted for evaluating federally funded programs throughout the United States. This framework was developed by a large group of evaluators and consultants in an effort to incorporate, integrate, and make accessible to public health practitioners useful concepts and evaluation procedures from a range of evaluation approaches. Using the computer in García's office, I quickly downloaded Figure 11.1 from the CDC Web site.

I then proceeded to describe the similarities of the three- and six-step approaches. I explained that the first two steps of the CDC framework (engage stakeholders and describe the program) corresponded to what I described as the first step of the Bunche–Da Vinci evaluation. The activities of CDC Step 3 (focus the evaluation design) are what we accomplish in the second step I described, and CDC Steps 4 to 6 (gather credible evidence, justify conclusions, and ensure use and lessons learned) correspond to what we achieve in Step 3 of the Bunche–Da Vinci evaluation. In addition, I explained how the standards for effective evaluation (utility, feasibility, propriety, and accuracy; Joint Committee on Standards for Educational Evaluation, 1994) and the American Evaluation Association's (2004) *Guiding Principles for Evaluators* (systematic inquiry, competence, integrity and/or honesty, respect for people, and responsibilities for general and public welfare) are realized using this evaluation approach. Well, that did it; García and Chase were exhausted. They asked me if I could meet with them again the next week to further discuss establishing an evaluation contract.

Meeting 2. I could tell a considerable amount of discussion had occurred since our initial meeting. Although it was clear they were eager to proceed, I could sense I was about to be bombarded with more questions. First, García wanted to know whom would be talking to the stakeholders and arranging meetings to discuss the program and the evaluation. My guess is she was concerned that I (a highly educated European American male) would be perceived as a threatening outsider and might not be the best choice for engaging her predominately Latino and African American students, parents, staff, and teachers. This gave me the opportunity to impress on her that personnel recruitment, selection, and management is one of the most critical components of conducting a successful evaluation. She seemed to get this point when she thought about it in terms of problems she has encountered running her school. I assured her that we would strive to assemble a highly competent and experienced team with a particular emphasis on making sure we have team members knowledgeable about the program, context, and the evaluation topics we pursue. We would also make sure that we hire team members that share key background characteristics such as ethnicity, culture, language, and sociocultural experiences, and who possessed the ability to understand and build trusting and productive relationships with the various stakeholder groups represented at the Bunche–Da Vinci Learning Academy. Furthermore, we would request funds to support hiring top-level experts to consult with us on topics we encounter that require highly specialized expertise. She very much liked the idea of supporting the assembly of a multicultural team as part of the evaluation contract.

Next, Chase wanted to know if there were any risks or common problems associated with engaging stakeholders. After reminding him of the potential benefits, I described some of the risks related to external evaluation in general, as well as to the evaluation plan I was proposing. For example, it is possible that various stakeholders (i.e., the Da Vinci Learning Corporation administration or staff) would refuse to participate, provide misleading information, or undermine the evaluation in other ways. The evaluation findings might deliver various types of bad news, including uncovering unprofessional or illegal activities, and result in serious consequences for some stakeholders. Precious time and resources that could be used to provide services to needy students could be wasted if the evaluation is not accurate, useful, and cost-effective (see Donaldson, 2001b; Donaldson et al., 2002, for more possible risks). Of course, I explained there were also serious risks associated with not evaluating at this point and that we would attempt to identify, manage, and prevent risks or negative consequences of our work every step of the way. He seemed pleasantly surprised that I was willing to discuss the dark side of external evaluation and was not just another evaluation salesperson.

After fielding a number of other good questions, concerns about budget and how much all this professional evaluation service would cost emerged in our discussion. I proposed to develop separate budgets for the conceptual work to be completed in Steps 1 and 2 and the empirical evaluation work to be completed in Step 3. That is, we would be willing to sign a contract that enabled us to complete the first two steps of developing program theory, and formulating and prioritizing evaluation questions. Based on the mutual satisfaction and agreement of both parties, we would sign a second contract to carry out the empirical work necessary to answer the evaluation questions that are determined to be of most importance to the stakeholders.

Our evaluation proposal was intended to be cost-effective and to potentially save both parties a considerable amount of time and resources. During the completion of the first contract, Bunche–Da Vinci stakeholders would be able to assess the effectiveness of the evaluation team in this context and determine how much time and resources they want to commit to empirical data collection and evaluation. This first contract would provide enough resources and stability for our DGHK evaluation team to explore fully and better understand the program, context, stakeholders, and design and data collection constraints before committing to a specific evaluation design and data collection plan.

García and Chase seemed enthusiastic about the plan. They were ready to draw up the first contract so we could get to work. However, it dawned on them that some of their key colleagues were still out of the loop. They began to discuss which one of them would announce and describe the evaluation to their colleagues. At that point, I offered to help. I suggested that they identify the leaders of the key stakeholder groups. After introducing and conveying their enthusiasm for the idea and the DGHK evaluation team (preferably in person or at least by phone, as opposed to e-mail), they would invite these leaders to an introductory meeting where the evaluation team would provide a brief overview of the evaluation plan and invite them to ask questions. García and Chase invited corporate, faculty, staff, parent, and student representatives to our next meeting to learn more about the evaluation plan.

I have tried to provide a realistic account of how I would attempt to negotiate an evaluation contract with these potential clients. As part of this dialogue, I have simulated the types of discussions and questions I commonly encounter in practice. I would assemble a multicultural team (drawing on existing DGHK Associates staff) to introduce the evaluation plan to the larger group of stakeholder leaders. The presentation would aim to be at about the same level as just mentioned, with some additional tailoring to reach and be sensitive to the audience.

EVALUATION PLAN

In this section, I add some flesh to the bones of the evaluation plan proposed. More specifically, I provide a brief rationale for each step, more details about the actions we take, and some examples of what might happen as a result of our actions at each step of the plan. To stay within the bounds of this hypothetical case and intellectual exercise, I thought it would be useful to use a format that provides readers a window on how I describe the Bunche–Da Vinci evaluation to prospective evaluation team members. Therefore, I strive to illustrate the level of discussion and amount of detail I typically provide to the candidates being interviewed for the DGHK Associates multicultural evaluation team. My goal is to illustrate how I would provide a realistic job preview to those interested in joining the team as a way to help readers gain a deeper understanding of my evaluation plan. Realistic job previews are popular human resource selection and organizational socialization interventions that involve explaining both desirable aspects of a job and potential challenges upfront, in an effort to improve person–job fit and performance and reduce employee dissatisfaction and turnover (Donaldson & Bligh, 2006).

Bunche–Da Vinci Realistic Job Preview

The evaluation of the Bunche–Da Vinci partnership uses a program theory-driven evaluation science framework. It emphasizes engaging relevant stakeholders from the outset to develop a common understanding of the program in context and to develop realistic expectations about evaluation. We accomplish this by tailoring the evaluation to meet agreed-on values and goals. That is, a well-developed conceptual framework (program theory) is developed and then used to tailor empirical evaluation work to answer as many key evaluation questions as possible within project resource and feasibility constraints. A special emphasis is placed on making sure the evaluation team members, program theory, evaluation questions, evaluation procedures, and measures are sensitive to the cultural differences that are likely to emerge in this evaluation.

Step 1: Developing Program Theory. Our first task is to talk to as many relevant stakeholders as possible to develop an understanding of how the Bunche–Da Vinci program is expected to meet the needs of its target population. For efficiency sake, we work with four or five groups of five to seven stakeholders' representatives to gain a common understanding of the purposes and details about the operations of the program. Specifically, you (interviewee) are asked to lead or be part of an interactive process that makes implicit stakeholder assumptions and understandings of the program explicit. (See Donaldson & Gooler, 2003, and Fitzpatrick, 2002, for a detailed discussion and examples of this interactive process applied to actual cases.)

Let me give you an example based on some of the characteristics and concerns I have learned so far. The Bunche–Da Vinci Learning Partnership Academy is an elementary school located in a tough neighborhood. It is a unique partnership between the school district and a nonprofit educational company specializing in innovative school interventions for low-performing students. The school population is characterized by transience, illegal enrollments originating from the adjacent district, high numbers of non-English-speaking students, high levels of poverty, a young and inexperienced staff with high turnover, and geographical isolation from the rest of the district. The principal and superintendent are concerned that the partnership program is not an effective way to educate their students. They have shared with me a number of hunches they have about why the program is not working and the principal has some ideas about how to change and improve the school. But it is important to keep in mind that as we engage other stakeholders in discussions about the program, we are likely to gain a

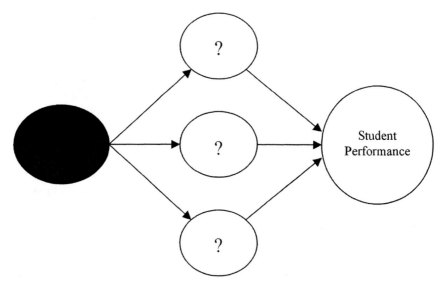

Figure 11.2.   Example of program impact theory.

wealth of additional information and possibly hear extremely different views about the program's success and challenges.

After we collect and process the information we gather in the stakeholder meetings, we attempt to isolate program components, expected short-term outcomes, more long-term or ultimately desired outcomes, and potential moderating factors (Donaldson, 2001a). For example, student performance on state test scores has been the main desired outcome discussed in the conversations I have had with the school administrators. Other stakeholders may strongly object to the notion that the program is designed to improve state test scores and be upset by the No Child Left Behind legislation and accountability zeitgeist. In fact, I expect they will provide us with a range of other desired outcomes to consider as we try to gain a deep understanding of the program. However, the program's impact on student performance will likely end up being one of the desired outcomes we explore conceptually and potentially evaluate. See Figure 11.2.

Figure 11.2 shows an example of how we begin to diagram and probe stakeholders' views about program impact on student performance. The anchor of this program impact theory is student performance. It assumes that the partnership program compared to no program or an alternative (e.g., a typical curriculum in a comparable school) is expected to improve student performance. Our discussion with the stakeholders attempts to clarify why or how the partnership program is presumed to accomplish this. We may discover that there are key short-term outcomes or mediating factors (represented by the question marks at this point) that are expected to result from the program, which in turn are expected to lead to improved student performance. Once we have clarified these expected mediating processes, we begin to probe whether these links are expected to be the same for all students and in all context variations that may exist across the delivery of the program. If not, we will isolate the key student characteristics (such as gender, ethnicity, socioeconomic status, language, acculturation, and attendance) and potential contextual factors (such as group or class dynamics, instructor effects, service-delivery characteristics, and the like) that could moderate or condition the strength or the direction of the arrows in our program impact theory. Our ultimate goal is to work through this interactive process with the diverse stakeholder groups until we have a common understanding about the purposes and expected benefits and outcomes of the program.

Once we have completed this process with the stakeholders, you and the other members of the team are required to assess the plausibility of the stakeholders' program theory or theories. You do this by reviewing the available research and evaluation literature related to factors identified. We specifically look for evidence that may suggest that some of the links are not plausible or that there may be side effects or unintended consequences we have not considered. The findings from the review, analysis, and team discussions may lead us to suggest that the stakeholders consider revising or making some additions to their program theory or theories. I expect it takes us at least 3 months of full-time work to complete this first step of our evaluation plan.

Step 2: Formulating and Prioritizing Evaluation Questions. Once we have a deep understanding of the program and context, we focus on illuminating empirical evaluation options for the stakeholders. You and the rest of the team are asked to frame potential evaluation questions so that they are as concrete and specific as possible and informed by the program theory or theories. The types of questions likely to be considered in the Bunche–Da Vinci evaluation fall under the categories of *program need, design, delivery* or *implementation, outcomes, cost,* and *efficiency* (see Rossi et al., 2004).

My best guess, based on what we know so far about the partnership program, is that we will most likely pursue questions about curriculum implementation, program operations and educational services delivery, and program outcomes. For example, the stakeholders may decide that they want empirical data to answer questions such as:

- Are administrative and educational service objectives being met?
- Are the intended curricula being delivered with high fidelity to the intended students?
- Are there students or families with unmet needs that the program is not reaching?
- Do sufficient numbers of students attend and complete the curriculum?
- Are teachers and students satisfied with the curriculum and educational services?
- Are administrative, organizational, and personnel functions of the partnership program effective?

Furthermore, questions about program outcomes will likely include:

- Are the desired short-term outcomes (mediators) being achieved?
- Are the desired longer term or ultimate outcomes of concern being achieved?
- Does the program have any adverse side effects?
- Are some recipients affected more by the program than others (moderator effects)?
- Does the program work better under some conditions than others (moderator effects)?

It is possible that we will be asked to pursue questions about the partnership program cost and efficiency—for example, (a) Are the resources being used efficiently? (b) Is the cost reasonable in relation to the benefits? (c) Would alternative educational approaches yield equivalent or more benefits at less cost? Furthermore, García does have some ideas for changing the program and may ask us to answer questions about students' needs and best ways to satisfy those needs. But I do think it is likely that you will pursue questions about program implementation and outcomes if both parties decide to enter into a second contract to collect data to answer the stakeholders' evaluation questions.

Once a wide range of potential evaluation questions has been formulated, you and the DGHK evaluation team will help the stakeholders prioritize the questions so that it is clear which questions are of most value. You need to note differences of opinion about the value of each question across the stakeholder groups and factor them into final decisions about which questions to pursue in the evaluation. In an ideal evaluation world, the entire range of relevant evaluation questions would be answered, and the program impact theory would be tested in the most rigorous fashion possible. However, in most evaluations, only a subset of questions and components of a program impact theory can be evaluated due to time, resources, and practical constraints. Prioritizing and identifying the most important evaluation questions can prevent paralysis in evaluation (i.e., deciding to wait or not to evaluate at all). Recognizing that some key questions can be addressed, even though some or many components of the program impact theory and other evaluation questions cannot be examined at this time, helps you to facilitate the evaluation process to move forward.

Finally, it is important for you to realize as a prospective employee and team member that the stakeholders may decide that they have learned enough about the program after we complete these first two steps. For example, it is not uncommon to discover that a program is obviously not being implemented as intended. This could lead the stakeholders to focus their attention and resources on fixing the program; or they may determine it is not repairable and decide to terminate the program and replace it with a promising alternative. If this type of situation develops, it is possible we would obtain an evaluation contract to help them develop and evaluate the new initiative, or we are likely to have another interesting evaluation contract in the firm that you could be hired to work on.

Step 3: Answering Evaluation Questions. Assuming the stakeholders decided they wanted to enter into the second contract with us to answer key evaluation questions, you would be asked to help design and work on an evaluation that would strive to answer those questions convincingly. In many respects, our evaluation approach is method neutral. We believe that quantitative, qualitative, or mixed methods designs are neither superior nor applicable in every evaluation situation (Chen, 1997; Reichhart & Rallis, 1994). Instead, our methodological choices in this evaluation are informed by (a) program theory, (b) the specific evaluation questions the stakeholders have ranked in order of priority, (c) validity and use concerns, and (d) resource and practical constraints (feasibility). Your main charge at this stage of the evaluation plan is to determine what type of evidence is desirable and obtainable, and to answer stakeholder questions of interest with an acceptable level of confidence.

As you might know, several factors typically interact to determine how to collect the evidence needed to answer the key evaluation questions such as stakeholder preferences, feasibility issues, resource constraints, and evaluation team expertise. Program theory-driven evaluation science is primarily concerned with making sure the data collection methods are systematic and rigorous, and produce accurate data, rather than privileging one method or data collection strategy over another (Donaldson & Christie, 2006).

It is typically highly desirable to design evaluations so that stakeholders agree up front that the design will produce credible results and answer the key evaluation questions. Participation and buy-in can increase the odds that stakeholders accept and use evaluation results that do not confirm their expectations or desires. Of course, making sure the design conforms to the Joint Committee Standards (Joint Committee on Standards for Education Evaluation, 1994) and the American Evaluation Association's *Guiding Principles for Evaluators* (2004) as much as possible is also helpful for establishing credibility, confidence, and use of the findings.

The challenge for us in the Bunche–Da Vinci Evaluation is to gain agreement among the diverse stakeholder groups about which questions to pursue and the types of evidence to gather to answer those key questions. For example, we should be able to gather data from students, teachers, school and corporate administration and staff, parents, and experts in areas of concern. Furthermore, if needed, it looks as if we will be able to collect and access data using existing performance measures and data sets, document and curriculum review, interview methods, Web-based and traditional survey methods, and possibly observational methods, focus groups, and expert analysis.

As a member of the evaluation team, your role at this stage of the process is to help facilitate discussions with the relevant stakeholders about the potential benefits and risks of using the range of data sources and methods available to answer each evaluation question of interest. You are required to educate the stakeholders about the likelihood that each method under consideration will produce accurate data. You need to discuss potential threats to validity and possible alternative explanations of findings (Shadish et al., 2002), the likelihood of obtaining accurate and useful data from the method in this specific situation, research with human participants (especially with minors), and informed-consent concerns, and to estimate the cost of obtaining data using each method under consideration. Once the stakeholders are fully informed, your job is to facilitate a discussion that leads to agreement about which sources and methods of data collection to use to answer the stakeholders' questions. This collaborative process also includes reaching agreement on criteria of merit (Scriven, 2003) or agreeing on what would constitute success or failure or a favorable or unfavorable outcome, which will help us justify evaluation conclusions and recommendations. If you are successful at fully informing and engaging the stakeholders, we believe it is much more likely that the stakeholders will accept, use, and disseminate the findings and lessons learned. If you allow me to make some assumptions, I can give you examples of how this third step of our evaluation plan could play out in the Bunche–Da Vinci evaluation you would be hired to work on.

First, I underscore that the information we have so far is from the school administrators. That is, once we hear from the teachers, parents, students, and corporate administration and staff, we may gain a very different account of the situation. In our view, it would be a fundamental error at this point to base the evaluation on this potentially limited perspective and not include the other stakeholders' views. Nevertheless, I make some assumptions based on their perspectives to illustrate the process that you will be asked to facilitate at this stage of the evaluation.

- *Assumption*: The stakeholders have decided that their top priority for the evaluation is to determine why the two separate indicators of student performance (state test scores; Da Vinci test scores) provide substantially different results.

García and Chase have suggested that their top concern is that student performance, particularly on the English-Language Arts components of state standards tests, have declined over the course of the partnership program. In fact, in the most recent testing (year 3), students scored lower than in years 1 and 2. These decreasing state test scores contrasted sharply with their corporate partner's assessments of students' performance. On the company's measures, the percentage of students reading at grade level had doubled over the past 3 years. Da Vinci staff from headquarters claimed to have heard students say that they were "finally able to read" and were much more enthusiastic learners. García and Chase are very concerned about this discrepancy.

- *Assumption*: We will assume that the stakeholders have produced Figure 11.2 and that increasing student performance is one of the main purposes of the partnership program.

In this case, we would explore all of the strengths and weaknesses of the feasible options for determining why measures of student performance are yielding different results. I would expect the stakeholders to decide to have us conduct a systematic and rigorous analysis of the construct validity of each set of measures. We would pay particularly close attention to potential differences in construct validity across our diverse and changing student body. In addition to the expertise on our team, we would likely hire top-level consultants with specific expertise in student performance measurement in similar urban school environments to help us shed light on the discrepancies. It will be critical at this stage of the evaluation to decide whether the performance problems are real or an artifact of inadequate measurement that can be explained by threats to construct validity (Shadish et al., 2002). Imagine the implications of the potential evaluation finding that performance is not really declining or if corporate measures are seriously (possibly intentionally) flawed.

- *Assumption*: The stakeholders decided that their second priority is to determine if the partnership program curriculum is being implemented as planned with high fidelity.

We will have learned much about the design of the curriculum and why it is believed to be better than the alternatives during Steps 1 and 2 of our evaluation process. It will be your job as a member of the evaluation team to help verify whether the curriculum is actually being implemented as intended. As is sometimes the case in educational settings, school administrators suspect the teachers might be the main problem. They have suggested to us that teachers have not bought into the partnership program, may even resent its requirements and added demands, and may just be going through the motions. They have also suggested to us the teachers are young and inexperienced and may not have the motivation, expertise, and support necessary to implement the program with high fidelity. Furthermore, there are some doubts about whether groups of students are fully participating in the program and adequately completing the lesson plans. The rather dramatic changes in the student population may have affected the quality of the implementation of the curriculum.

After considering the available data collection options for answering this question, I would expect you and the evaluation team to be asked to observe and interview representative samples of teachers about the delivery of the Bunche–Da Vinci curriculum. You could also be asked to gather data from students and parents about their experiences with the curriculum. For these technology-enriched students, I would expect that a Web-based survey could be designed and completed by many of the students. However, alternative measurement procedures would need to be developed for those too young to complete a Web-based survey. Furthermore, a representative sample of students could be interviewed in more depth. For the parents, I would expect we would need to design a standard paper-and-pencil survey, and be able to interview and possibly conduct some focus groups to ascertain their views and experiences. Finally, interviews of key staff members of both the school district and corporate partner would be pursued to further develop our understanding of how well the curriculum has been implemented and how to improve implementation moving forward. Keep in mind that even if we do find valid student performance indicators in the previous analysis, they could be meaningless in terms of evaluating the partnership program if the program has not been implemented with high fidelity.

The final example of a question that could be pursued in Step 3 of the Bunche–Da Vinci evaluation focuses on determining whether desired short-term outcomes have resulted from the program. For this example, I assume that the partnership program has been implemented with high fidelity. Figure 11.2 illustrates that three main short-term outcomes are expected to result from the partnership program. Let us assume that the stakeholders have agreed that the first one is a high level of intrinsic engagement of the curriculum. That is, the program produces a high level of intrinsic engagement, which in turn is expected to lead to increases in student performance.

- *Assumption*: The stakeholders decided their third priority is to assess whether Bunche–Da Vinci students have a high level of intrinsic engagement.

Imagine that after weighing the options, the stakeholders have decided that they would like us to measure and determine whether students at Bunche–Da Vinci have a high level of intrinsic engagement with the curriculum. You and the team would be asked to work with the stakeholders to develop a clear understanding and definition of this construct. Next, we search and critically review the literature to determine if there are good measures of this construct that could be used for this purpose. Assuming we have identified a strong measurement instrument (we would need to create one otherwise), we then develop another set of measurement procedures for the students to complete (or include the items on the previous instrument used, if possible). We would also make sure that we included items about key student characteristics of concern (such as ethnicity, language, acculturation, family, attitudes toward technology, and the like) and characteristics of the context (e.g., peer group dynamics and out-of-class study environment), exploring if there may be important moderating influences on the link between the program and this short-term outcome. This allows us to do more finely grained analyses to estimate whether the program is affecting this short-term outcome more for some students than others.

We assume that for this initial examination of intrinsic engagement, we do not have a reasonable comparison group or baseline data available. Therefore, prior to implementation, we gain agreement with the stakeholders about how we define high versus low levels of intrinsic engagement (i.e., establish criteria of merit). Finally, against our recommendation, we assume that the stakeholders have decided not to survey or interview parents or teachers about intrinsic engagement due to their lack of enthusiasm about spending additional resources for their third priority evaluation question.

As I hope you can now appreciate, working on the DGHK evaluation of Bunche–Da Vinci promises to be a meaningful opportunity for you. You will be part of a multicultural team of evaluation professionals engaged in helping to address a socially important set of concerns. The course of these children's educational careers and lives could be undermined if we find that this situation is as bad as it appears on the surface and sound recommendations for improvement are not found and implemented in the near future. I would now like to hear more about why your background, skills, and career aspirations make you a strong candidate for being an effective member of the DGHK/Bunche–Da Vinci evaluation team. But first, do you have any further questions about the job requirements?

REFLECTIONS AND CONCLUSIONS

The Bunche–Da Vinci Case presented DGHK Evaluation Associates with a challenging mystery to be solved. A highly complex set of potentially interactive factors appears to be suspect in the apparent demise of an innovative partnership program. Whomever or whatever is the culprit in this case seems to be responsible for undermining the performance of a diverse and disadvantaged group of students. In the face of this complexity, DGHK Evaluation Associates has proposed to use a relatively straightforward three-step process to develop and conduct evaluation services to help crack the Da Vinci Code and to potentially improve the lives and trajectories of these children. The proposed evaluation approach is designed to provide cost-effective, external evaluation services. DGHK Evaluation Associates promises to strive to provide evaluation services that are as accurate and useful as possible to the Bunche–Da Vinci stakeholders, as well as to work in a manner that is participatory, inclusive, and as empowering as stakeholders and constraints will permit.

In an effort to achieve these promises, I proposed to tailor the evaluation to contingencies our team encounters as they engage stakeholders in the evaluation process. Obviously, to complete this exercise of describing how program theory-driven evaluation science could be adapted and applied to this hypothetical case, I had to make many assumptions. Examples of the details of each step could be substantially different in practice if different assumptions were made. For example, if I assumed the stakeholders wanted us to propose how we would determine the impact of the program on student performance outcomes using a rigorous and resource-intensive randomized controlled trial (or quasi-experimental design, longitudinal measurement study, intensive case study, or something else), the particulars of the three steps would differ substantially. However, it is important to emphasize that the overall evaluation plan and process I proposed would be virtually the same.

The sample dialogue with the school administrators during the contracting phase and in the realistic job preview I gave to the potential evaluation team members were intended to be helpful for understanding how I provide evaluation services in real-world settings. Based on the case description, I predicted this evaluation would need to operate under somewhat tight resource and practical constraints and would be likely to uncover intense conflicts and dynamics among stakeholder groups. I tried to underscore the point that a fatal flaw would have been to design an evaluation plan in response to information and views provided almost entirely by one powerful stakeholder group (the school administrators, in this case). It seemed likely that the teachers and teachers' union may have made some different attributions (e.g., management, leadership, and organizational problems) about the long list of problems and concerns the administrators attributed to the "young and inexperienced" teachers. It also seemed likely that the corporate leadership and staff could have a very different take on the situation. Based on my experience, I am confident that the failure to incorporate these types of stakeholder dynamics in the evaluation plan and process would likely undermine the possibility of DGHK Evaluation Associates producing an accurate and useful evaluation for the Bunche–Da Vinci learning partnership.

It would have also been problematic to conduct extensive (and expensive) data collection under the assumption that student performance had actually declined. That is, a considerable amount of evaluation time and resources could have been expended on answering questions related to why performance had declined over time, when in fact performance was not declining or even improving, as one of the indicators suggested. Therefore, in this case, it seemed crucial to resolve the discrepancy between the performance indicators before pursuing evaluation questions based on the assumption that performance had actually declined.

Due to space limitations, there are aspects of this case and evaluation plan I was not able to explore or elaborate on in much detail. For example, during the developing program theory phase of the evaluation process, we would have explored in detail the content of the innovative technology-enriched curriculum and its relevance to the needs of the culturally diverse and changing student population. During Step 3, we would have facilitated discussions with the stakeholders to determine how best to disseminate evaluation findings and the lessons learned from the Bunche–Da Vinci evaluation. Furthermore, we would have explored the potential benefits and costs of spending additional resources on hiring another evaluation team to conduct a metaevaluation of our work.

In the end, I must admit I encountered strong mixed emotions as I worked on this hypothetical case and evaluation plan. As I allowed my imagination to explore fully the context and lives of these students and families, I quickly felt sad and depressed about their conditions and potential plight, but passionate about the need for and opportunity to provide help and external evaluation. As I allowed myself to imagine what could be done using external evaluation if there were no time, resource, and practical constraints, I became elated and appreciative about being trained in evaluation and inspired to apply evaluation as widely as possible. However, this was quickly dampened when I realized I have never encountered a real case in 20 years of practice without time, resource, and practical constraints. My spirits were lowered even more when I imagined the risk of using scarce resources to pay the salaries and expenses of well-educated professionals to provide unnecessary or ineffective evaluation services, when these resources would otherwise be used to educate and help these at-risk students and families. Of course, the beauty of this exercise, just like in a nightmare, is that I would quickly elevate my mood by reminding myself I am dreaming. Now that I (and my colleagues in this volume) have walked this imaginary tightrope with you, I hope you have a better understanding of the value, challenges, and risks of external evaluation. I imagine I do.

LESSONS LEARNED AND FUTURE DIRECTIONS

12

Lessons Learned From Applications of Program Theory-Driven Evaluation Science

The purpose of this chapter is to highlight some of the lessons that have been learned from the applications of program theory-driven evaluation science. The evaluations presented in part II illustrate in great detail how the dynamic, three-step process can play out in "real world" evaluation settings. Each of the cases raises different challenges for the evaluators and stakeholders, and it should now be apparent that stakeholder decisions about evaluation questions, design, methods, and contract requirements vary considerably across applications.

The dynamic interplay between evaluators and stakeholders and the great variability in evaluation procedures across applications underscores a main point of this book, which is that observing and understanding what takes place in actual evaluation practice is essential for advancing the understanding of the profession and discipline. That is, most evaluation texts and writings present evaluation theory (or theories) that prescribes how evaluation should be done in practice. Very few of these prescriptions are based on descriptive evidence. It has been increasingly suggested in recent years that the advancement of evaluation theory and practice relies on our ability to move from the prescriptions of how to practice toward developing a research base to guide practice (Henry & Mark, 2003; Donaldson & Lipsey, 2006). Of course, there are a number of ways for expanding the evidence base related to how best to practice evaluation. For example, there is now research on how evaluation practitioners use evaluation theory (Christie, 2003); how evaluation theorists actually practice evaluation (Alkin & Christie, 2005); stakeholder participation dynamics (Campbell & Mark, 2006); evaluation utilization (Cousins & Leithwood, 1986); stakeholder bias (House, 2003);

strategies for preventing excessive evaluation anxiety (Donaldson, Gooler, & Scriven, 2002); as well as forthcoming descriptions and analyses of how particular evaluation approaches or theories of practice evolve over time in exemplary applications (see Fitzpatrick, Mark, & Christie, in press).

In this chapter, I reflect on observations of evaluators and stakeholders working together across the three steps of the program theory-driven evaluation process in the Work and Health Initiative (WHI) evaluation. Specifically, I discuss each program evaluation and highlight lessons that might be gleaned from the three steps: (a) developing program impact theory; (b) formulating and prioritizing evaluation questions; and (c) answering evaluation questions. I also capture some of the insights, experiences, and practical advice that might be culled from across the evaluations within the WHI. Next, I suggest what we might take away from the exercise of using program-theory driven evaluation to propose evaluation services to the Bunche–Da Vinci Learning Partnership Academy stakeholders. Finally, I draw links between applications of program-theory driven evaluation science in the effectiveness evaluation domain and its use in the efficacy evaluation literature. These observations and reflections are made in the spirit of learning how to better provide evaluation services that promote the positive development of people, programs, and organizations.

EVALUATION OF WINNING NEW JOBS

The application of program theory-driven evaluation science for developing and improving a reemployment program and three organizations funded to implement the program to unemployed workers in the State of California was presented in detail in chapter 5. Three Winning New Jobs (WNJ) sites served 5,290 unemployed or underemployed Californians in 455 WNJ workshops over a 4-year period. WNJ program sites served a very diverse population of Californians in terms of ethnicity, age, educational background, and employment history. Nearly two thirds (60%) of WNJ participants were female; more than two thirds (68.7%) were people of color; more than two fifths were over 40 years of age (41%); and nearly half held a high school degree or had fewer years of formal education (44.8%).

A parsimonious program impact theory was used to focus the WNJ evaluation on questions that fell within five main areas: program implementation, program service, short-term outcomes, reemployment outcomes, and program sustainability and replication. Using rather extensive quantitative and qualitative databases, findings and evaluative conclusions were presented to illustrate the actual products of this program theory-driven evaluation. I now explore some of the benefits and challenges of using program theory-driven evaluation in this specific case.

Developing Program Impact Theory for WNJ

The WNJ program was based on evidence from a relatively strong tradition of evaluation and applied research (Price et al., 1992; Vinokur et al., 1991). Discussions of program impact theory with stakeholders quickly became focused on the social science theory and empirical support of the program from prior research. The program designers were very clear about what they were attempting to achieve in this project. They convinced the stakeholders that they had a strong empirical base of evidence to suggest that the core program impact theory would be effective for California workers. It was believed that the fundamental challenge was to make sure the WNJ program was implemented with high fidelity to the original model, with only minor adaptations to local conditions. A plausibility check with literature, and extensive probing of the arrows in the model by the evaluation team, suggested that the program might work better for some Californians than others (e.g., those who needed the core proximal outcomes in order to find a good job).

The WNJ program impact theory was developed rather quickly and easily, and was generally well accepted by stakeholders across the project (see Fig. 12.1). It was reported often that the WNJ program impact theory was very helpful for organizing project management, training project staff, and for describing the program to those unfamiliar with it. However, as the project unfolded, it became clear that there were a number of moderating factors believed to be at play in the application of this program in California. Although it was decided to retain the core program impact theory as a way of communicating the purpose of the program, a number of potential moderators were analyzed throughout the evaluation.

The WNJ program impact theory suggested that participants would receive mental health gains as a result of completing the program. However, the evaluation team was not able to overcome stakeholder fears about implementing a data collection strategy that would assess changes in mental health. In a nutshell, the program staff believed that asking mental health questions would offend participants, and would likely undermine the success of the program. This problem limited the evaluation to examining only one of the main desired outcomes of the program—reemployment.

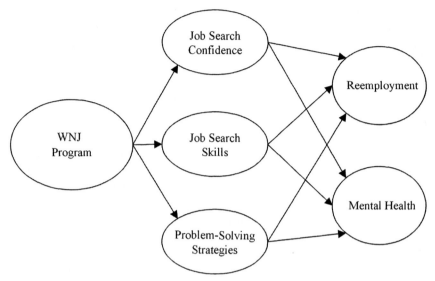

Figure 12.1.   WNJ program impact theory.

Formulating and Prioritizing Evaluation Questions for WNJ

Formulating possible evaluation questions came easily to the WNJ stakeholders. There was great enthusiasm from the start about the potential for the evaluation to advance the knowledge and understanding of the WNJ program. However, this enthusiasm seemed to dampen when the discussion turned to prioritizing the evaluation questions. The prioritization stage exposed the fact that evaluation resources were limited in a way that some of the key outcome questions would not be answered with "a high level of confidence" and others would not be answered at all. The fundamental tension here was over the evaluation standard of feasibility and the scope of obligation of the implementing organizations on one hand, and the desire for an unequivocal impact assessment by the program design and evaluation teams on the other hand (i.e., focus on the accuracy standard of evaluation; see Donaldson & Lipsey, 2006).

The evaluation team organized a number of meetings and conference calls to fully explore stakeholder differences and to find a compromise position. This prioritization phase of the program theory-driven evaluation science process was by far the most difficult and time-consuming aspect of designing the evaluation of the WNJ program. In the end, The

California Wellness Foundation (TCWF) decided that the feasibility and scope concerns were legitimate, but seeking additional evaluation resources to rigorously examine the impact of WNJ on reemployment and health outcomes was beyond what they desired to accomplish. This decision by the project sponsor broke the deadlock and led the stakeholders to focus on the five categories of evaluation questions discussed previously. It is important to point out that the failure to reach agreement about evaluation question priorities and what constitutes credible evidence seemed to haunt the rest of the project and evaluation. That is, there seemed to be less camaraderie and enthusiasm across the stakeholder groups at times, much more tension and competition in meetings, and more concern about evaluation procedures and findings.

Answering Evaluation Questions for WNJ

The data collected to inform program implementation and service seemed to be highly valued and regularly used to improve the work of program staff and management. Regular evaluation reports and meetings with each organization helped guide and improve the implementation of WNJ at all three sites. There was strong agreement across all the stakeholder groups that findings from the formative and summative evaluation of WNJ implementation and service were a critical component in the success of the WNJ project.

Many of the stakeholder groups also seemed to value, trust, and use the evidence to make changes in the short-term outcomes (i.e., improved job-search confidence, job-search skills, and problem-solving strategies, as well as the longer term reemployment outcome data). Although some of these findings remained equivocal because of noted evaluation design constraints, the overall pattern of findings from the evaluation suggested that WNJ provided some help for displaced California workers.

The lack of evidence for the WNJ program's sustainability and replication seemed to indicate that much more work needed to be done to reach these project goals. The lack of unequivocal evaluation evidence demonstrating that WNJ is superior to existing programs in the state undoubtedly limited the ability of the WNJ team to make additional progress on facilitating the adoption of WNJ throughout the state of California.

EVALUATION OF COMPUTERS IN OUR FUTURE

The Computers In Our Future (CIOF) project created 14 community computing centers (CCCs) in 11 low-income California communities. The 14 CCCs were designed to demonstrate innovative, creative, and culturally sensitive strategies for using computer technology to meet the economic, educational, and developmental needs of their local communities. The CIOF program explored and demonstrated ways in which CCCs can prepare youth and young adults ages 14 through 23 to use computers to improve their educational and employment opportunities, thereby improving the health and well-being of themselves, their families, and their communities.

Organizational readiness criteria were used to select 11 diverse California organizations to accomplish these goals. These organizations were diverse with respect to organizational type, geographical location, and populations served. Collectively, these centers were designed to provide access to more than 200 computer workstations statewide. With respect to open access service goals, each site collectively committed to providing unrestricted open access to 27,705 Californians (approximately 6,900 individuals per year, statewide). Similarly, they committed to providing technology training and work experiences to over 4,300 youth and young adults over the 4-year program period.

Developing Program Impact Theory for CIOF

The CIOF program impact theory was quite challenging to develop. Each CCC was charged with developing a program that targeted the needs of their surrounding community. Initial discussions with CCC leaders and the CIOF program coordination team suggested that it may be difficult to develop one overarching program impact theory that would be useful for all 11 organizations. Many of the initial ideas about the program model were not consistent with prior research and theory about related topics. That is, the plausibility check performed by the evaluators helped considerably to shape the process. After much discussion and negotiation over a 12-month period with extensive input from site leaders, the funder, and the CIOF program coordination team, stakeholders were able to reach agreement on the program impact theory presented in Figure 12.2. Figure 12.2 shows that participation in the CIOF program was believed to lead to improved attitudes toward computers, technology skills, career development knowledge, job-search skills, and basic life skills. These acquired skills and knowledge were presumed to facilitate the pursuit of more education, internship opportunities, and better employment options, which, in the long term, is expected to improve participants' health status. Many stakeholders reported that the program impact theory helped them better conceptualize, design, and deliver program services that were likely to affect the desired target outcomes.

One of the conceptual challenges encountered in the CIOF project was making a plausible link to the WHI goal of improving health status. As we pushed the stakeholders to articulate the nature of paths leading

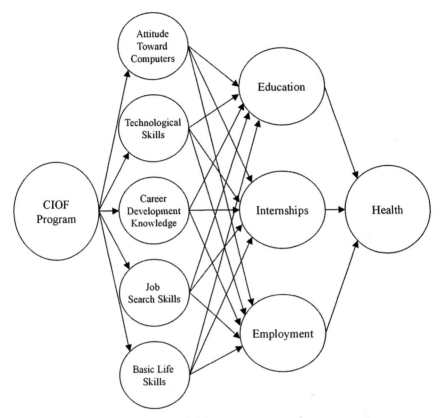

Figure 12.2.   CIOF program impact theory.

to health, it became clear that health was conceptualized as a rather distal outcome of CIOF. For the purposes of the evaluation, the stakeholders chose to focus on the proximal and intermediate outcomes. The link between education, internships, employment, and health were assumed to be long term, so they were not examined empirically in the evaluation. On balance, the program impact theory for CIOF proved to be a useful tool for discussing each centers' goals and aspirations, monitoring progress, and for helping stakeholders formulate and prioritize evaluation questions.

Formulating and Prioritizing Evaluation Questions for CIOF

After a rather lengthy process of developing a program impact theory for CIOF, the evaluation team and stakeholders explored a wide range of potential evaluation questions. Using the networks and communication channels established for developing program impact theory, the evaluation team facilitated a discussion to prioritize evaluation questions. Somewhat surprisingly, the stakeholders seemed to reach agreement about the key evaluation questions much more quickly than they did about the program impact theory. It was decided that evaluation resources being allocated to CIOF would be deployed to answer the following questions:

- What does it take to set up a vibrant, accessible, relevant, and sustainable community computing center?
- What does implementation look like?
- What are the key challenges and success factors to program development and implementation?
- Whom are the sites serving (e.g., population characteristics of the service recipients)?
- How many people are served at each site?
- How do technology access and training improve the employment prospects of young people?
- What are realistic outcomes in increased computer or technical skills, employment, increased literacy, English acquisition, attainment of GED, or other educational targets?
- How does the center affect the participants personally (e.g., self-confidence, motivation, life skills, etc.)?
- What are the demonstrable payoffs to communities (e.g., increased cohesion, access to technology resources and services, etc.)?
- What are the strengths and weaknesses of the specific CIOF models at each of the CCCs (i.e., access, education, resource, and voice)?

This set of evaluation questions was highly valued by the stakeholders and seemed to be within the range of what the evaluation team believed could be answered within the resource constraints of the project. An important lesson illustrated at this step was how successful relations and communication patterns established during the previous step can be useful for facilitating subsequent steps in the process, which is particularly important for projects as complex as CIOF.

Answering Evaluation Questions

The data collected to answer the CIOF questions involved the extensive use of quantitative and qualitative methods. Data were collected for over 25,000 program participants, including (a) user data (demographic and background information) for 22,729 individuals, (b) center-usage data (daily activities and time spent at the center) for 12,049 individuals,

(c) pretest and posttest data for over 400 individuals, and (d) follow-up interview data for over 200 individuals. Both user and usage data were collected for nearly half (47%) of all individuals tracked. Data collected with these measures included demographic and background data, computer-related behaviors inside and outside the CIOF centers, attitudes toward computers, computer skills, reactions to the CIOF centers, and educational and employment outcomes. These data were limited in similar ways to the WNJ data (e.g., one group, pretest–posttest research design) when they are used to estimate program impact. However, various types of qualitative implementation and outcome data were also collected, including site visit observations and interview data from site leaders and program coordination team members.

The fairly rapid feedback process to the stakeholders was particularly notable in the CIOF evaluation. That is, to support continuous program improvement within the CIOF implementation sites throughout the life of the initiative, the evaluation team prepared and disseminated 113 evaluation reports to CIOF program leaders, PCT members, and TCWF over the 5-year funding period. These included five year-end evaluation reports, four midyear evaluation reports, 92 site reports, four interim reports, and eight miscellaneous evaluation reports to program grantees. As tools for program monitoring and improvement, these reports documented not only key accomplishments and program activities, but also key program challenges and recommendations for addressing challenges. Conference calls and/or face-to-face meetings were held with site leaders, the PCT, and TCWF to discuss each report. In addition, the evaluation team presented key findings and updates on statewide service statistics at each of the biannual CIOF statewide conferences. During these various communications, the evaluation team facilitated discussion on the meaning of the findings and on developing strategies and responses to addressing program recommendations.

The communication and reporting processes in the CIOF project were overwhelming at times. We found that conference calls and/or face-to-face meetings were much more effective than the written reports for helping the stakeholders understand how best to use the evaluation findings. A main lesson learned from the CIOF project was to carefully negotiate upfront, and to continually examine and revise requirements for how data are to be reported to the stakeholders in a timely and meaningful fashion. The burden of heavy written reports was too great at times, and on reflection, the evaluation team believed they could have been more efficient and effective at ensuring use and sharing the lessons learned by relying less on written reporting formats.

EVALUATION OF THE HEALTH INSURANCE POLICY PROGRAM

The Health Insurance Policy Program (HIPP) was designed to support the development of state policy to increase access to health insurance for employees and their dependents that is not only affordable but is also comprehensive and emphasizes the promotion of health and prevention of disease. To this end, the HIPP issued an annual report on the state of health insurance in California based on surveys of the nonelderly population, HMOs, licensed health insurance carriers, purchasing groups, and employers. In addition, HIPP team members developed policy briefs and related health insurance publications for broad dissemination to appropriate policy stakeholders.

Developing Program Impact Theory for HIPP

During the first year of the project, the evaluation team began working with the project leaders and staff from TCWF to develop a program impact theory for the HIPP program. Discussions about program theory occurred over a 12-month period at face-to-face meetings, during conference calls, over e-mail, and through written document exchanges. There were several misunderstandings and disagreements early on about what the project leaders were and were not responsible for within the scope of their grant from TCWF. The process of developing program impact theory provided a valuable avenue for clarifying and revising expectations about the scope of the work.

As shown in Figure 12.3, the HIPP program sought to increase target constituents' awareness and understanding of the status of health insurance issues in California, and to influence policy development. The program theory shows that a range of publication development, report dissemination activities, and follow-up activities were going to be conducted in an effort to reach those desired outcomes. Support activities and potential outcomes are shown in dotted-line boxes to indicate that they are expected to occur, but are not required by the funding agency. This representation of program theory, using rectangles with bullets and the dotted-line rectangle, was strongly preferred by the stakeholders over other more typical representations (e.g., circles and individual arrows like the previous two cases). We found that being flexible about the format of the program impact theory was key to gaining buy-in for using program theory-driven evaluation to evaluate a research program of this nature.

Formulating and Prioritizing Evaluation Questions for HIPP

In collaboration with TCWF, the evaluation team led discussions about potential evaluation questions with the two project teams. After exploring

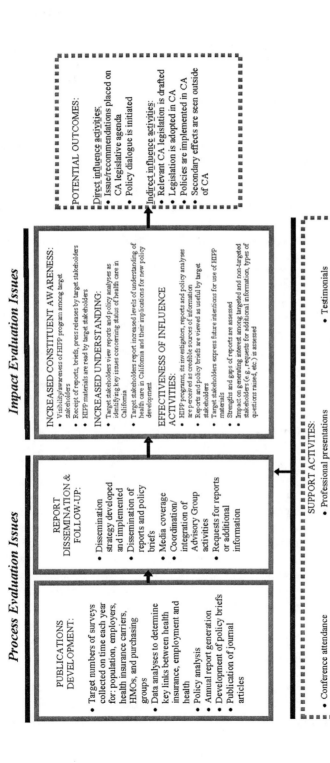

Figure 12.3.   HIPP program impact theory.

a wide range of possible evaluation questions, a discussion of which questions were most important to answer was facilitated by the evaluation team. The stakeholders eventually decided to focus the HIPP evaluation on the following questions:

- Did HIPP conduct surveys of California individuals, HMOs, insurers, employers, and purchasing groups?
- Were data analyzed and policy analysis conducted on the basis of data gathered?
- Were annual reports and policy briefs on the state of health insurance in California developed and disseminated to relevant policymakers?
- Did HIPP identify key trends and make relevant policy recommendations regarding California health care needs, health status, risk status, and health risk behaviors?
- Did HIPP identify access barriers to affordable health care, health promotion, and disease prevention?
- Were findings from different surveys integrated to impact policy development and target outcomes in meaningful ways?
- In what ways was HIPP effective in raising awareness among policymakers and the public about the status of health insurance in California and influencing the direction of health care policy?

There seemed to be much more concern expressed in the deliberations about final evaluation questions to pursue in HIPP than in some of the other evaluations. The nature of the contractual relationships between partners seemed to suggest that evaluation questions were limited to what was written in the original proposal. The highly participatory process used to generate and prioritize evaluation questions was only welcomed to a certain extent. That is, the evaluation team was required to keep in mind what was agreed to in the original contracts and to be cognizant that there were not additional resources to answer new questions without eliminating original questions or seeking new evaluation funding. The evaluation team was reminded how important it is to understand and acknowledge stakeholder expectations and contractual obligations related to workload when deciding on how many and which evaluation questions to pursue.

Answering Evaluation Questions

Interviews were conducted each year with a stratified, random sample of target audiences to address the evaluation questions. Key informants were stratified by organization type, including California policymakers, health insurers, HMOs, interest and advocacy groups, foundations, media, and university constituents. In-depth qualitative and quantitative data were gathered through face-to-face or telephone interviews to assess respondents' awareness of the research, understanding of material, how informative it was, and the influence of the research (i.e., credible, useful). The evaluation team also examined whether and how respondents used the HIPP research in their work. In addition, media coverage and direct and indirect policy changes were tracked. Finally, a random sample of individuals publishing in the health insurance arena over the past 5 years were selected to provide a critical peer review of the HIPP research. Overall, extensive qualitative and quantitative data were collected from over 300 key health care constituents over the program period, including three peer reviewers across the country not directly involved with the HIPP program.

The HIPP stakeholders seemed to warm up to the notion of using evaluation to verify the effectiveness of their actions and to facilitate the development of the HIPP program as the project matured. That is, after they experienced an improvement cycle using evaluation feedback, they seemed much more accepting and willing to engage in the requirement of evaluation. To support continuous program improvement within the HIPP research program, the evaluation team prepared and disseminated 13 evaluation reports to HIPP program leaders over the 4-year funding period. As tools for program monitoring and improvement, these reports documented not only key accomplishments and program activities, but also key program challenges and recommendations for addressing challenges. Conference calls and/or face-to-face meetings were held with program managers and TCWF program officers to discuss each report. During these various communications, the evaluation team facilitated discussion on the meaning of the findings and on developing strategies and responses to address recommendations. In addition, the evaluation team attended annual advisory group meetings and addressed questions pertaining to evaluation findings at those meetings. Overall, the formative evaluation processes specified here became viewed as a critical component for achieving the HIPP project goals.

EVALUATION OF THE FUTURE OF
WORK AND HEALTH PROGRAM

The evaluation of the Future of Work and Health (FWH) program provides an illustration of how evaluations are affected when the program design changes during the evaluation. In this case, there was considerable turmoil between stakeholders in terms of what should be the focus and scope of the project. Initial discussions between the stakeholders and evaluation team about developing a program impact theory quickly revealed that there was a lack of clarity about how TCWF funding was going to be used to support the program aim of understanding the rapidly changing nature of work and its effects on the health of Californians.

The program was reconfigured twice during the evaluation contract. Each time the evaluation team had to develop a new program impact theory to clarify the program's purpose and goals. The personnel involved with running the program also changed. Consequently, the evaluation team and stakeholders worked together for several years to finalize a program impact theory that could be used to generate evaluation questions. Additional delays, related to TCWF funding new projects to meet the new goals, further limited the empirical component of the evaluation.

Although the discussions of program impact theory and the evaluation questions were participatory and highly interactive, it was challenging to reach consensus on what and how to evaluate the FWH program. After much discussion and a fair amount of conflict, the relevant stakeholders agreed to the program impact theory shown in Figure 12.4, and to address questions about the trends identified, building a network, building knowledge, identifying policy, and dissemination. Document analysis and interviews were the primary forms of data collection used to answer the evaluation questions.

The description of the FWH evaluation in chapter 8 revealed how the three-step program theory-driven evaluation process held up under difficult circumstances. It could be argued that requiring the stakeholders to complete Step 1 before moving on to Steps 2 and 3 forced them to face difficult conceptual and personnel issues from the outset. Moving forward without gaining conceptual clarity about the program design would have likely led to ineffective programming and evaluation. Despite all of the trials and tribulations of implementing and evaluating the FWH program, the formative and summative evaluation processes appeared to help salvage what could have been a failed effort and waste of resources for TCWF. In the end, evaluation findings suggested that the FWH program was eventually able to reach many of its revised goals and objectives.

LESSONS LEARNED ACROSS THE EVALUATIONS OF THE WHI

This section is intended to capture some of the insights, experiences, and practical advice that can be learned about the program theory-driven approach to program planning and evaluation in the context of

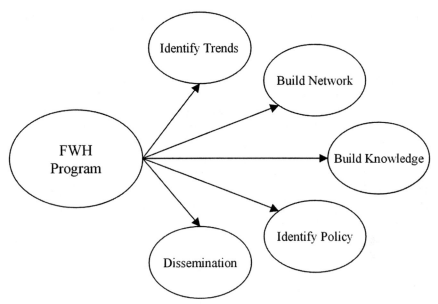

Figure 12.4.   FWH program impact theory.

this initiative (see Donaldson & Gooler, 2002). Although some of these lessons may be primarily relevant to conducting program theory-driven evaluations (e.g., lessons pertaining to developing program theory), we acknowledge others may extend to evaluation practice more generally (e.g., strategies for managing evaluation anxiety, providing continuous improvement feedback, building program and evaluation capacity, and managing the formative to summative evaluation transition). However, our focus here is to present lessons learned from the evaluation of the WHI that may be useful for refining some of the initial theorizing about how best to conduct program theory-driven evaluations, as well as improving program theory-driven evaluation practice.

Interactive, Bidirectional Program Impact Theory Development Was Most Effective

There seems to be a natural tendency to develop program theory moving from the left of a model to the right (as we were trained to read). This can be very limiting because most of the time with stakeholders is then spent talking about the program, and the etiology of outcomes and potential mediators and moderators are given short shrift. Another main consequence of the standard, linear, "left to right" approach is that stakeholders typically generate a long list of desired program outcomes, many of which appear unrealistic and only remotely connected to the program.

We have found that starting with desired outcomes is usually a more productive first step. Once we have identified a relatively short list of the outcomes the program is supposed to achieve, we begin the discussion of how this might occur. This leads us into the details of the how the program is expected to affect intermediate outcomes or mediators. Once we develop a first draft of a program impact theory based on discussions with stakeholders, we examine the various links in light of prior theory, research, and evaluation findings. This often leads us to suggest substantial revisions to the program impact theory or even to the program itself when it is highly inconsistent with what is known about the topic. The point here is that this process seems to work best when evaluators use a highly interactive, bidirectional approach to developing program theory (cf. Donaldson, 2001a, 2003; Fitzpatrick, 2002). This approach involves moving back and forth between stakeholder discussions and the literature several times, moving right to left and left to right and back again when thinking about the program impact theory and its various links between constructs, thinking in detail about the nature of each arrow (e.g., program effect decay functions), and carefully considering the characteristics of the participants, service providers, settings and the like that might affect (moderate) the relationships suggested by the program impact theory (see Donaldson, 2001a).

Reaching Consensus Was NOT Difficult

One concern about developing program impact theory in a collaborative fashion is that it will be difficult, if not impossible, to reach consensus among stakeholders. It is common to find that stakeholders initially have very different views and conceptions of their program. However, my experience in this initiative, as well as in other evaluations, is that the interactive process described usually leads to a high level of agreement and common understanding. If there are competing views of a program's impact theory, these views can usually be tested in the evaluation. As is the case in basic science, competing theoretical perspectives makes the empirical confirmation and/or disconfirmation (evaluation) process even more informative.

Complexity Is the Enemy: Strive for Parsimonious Program Impact Theories

We have found that there seems to be a natural tendency for stakeholders to represent all possible program components, mediators, moderators, outcomes, and the like in their program theories. However, overly complex program theories often confuse stakeholders and hinder discussions of the important evaluation questions that need to be answered (see Fitzpatrick, 2002). The interactive, bidirectional approach to program impact theory development described here strives to produce parsimonious program theories that can be used to develop a common understanding of the program among stakeholders and to help identify and prioritize evaluation questions (Donaldson, 2001a, 2003). Although I believe it is important to acknowledge that some programs and their effects may be very complex (see Donaldson, 2001a), at least a version of a program impact theory should solely focus on the main pathways between a program and its presumed outcomes. The resulting program impact theories presented in chapters 4 through 8 represent the most parsimonious versions of each program theory used to describe the programs in the WHI. As can be observed, the specification of program impact theory tends to vary across different evaluation contexts. That is, the level of complexity and type of model and/or display vary based on program type and stakeholder preference. It should be noted that the evaluators often discuss the program impact theory in much more detail when deciding how to design an evaluation and collect data to answer key evaluation questions (Donaldson, 2003).

Unpack Program Process Theory Selectively

Program process theory can be very useful in program planning and for understanding if a program is being implemented as intended. It is also useful for guiding evaluation questions about the implementation of a program. However, developing program process theory side-by-side with program impact theory usually overwhelms stakeholders. It is typically too confusing to link all of the components of a program process theory with the components of a program impact theory in one diagram. This is why I recommend keeping stakeholders focused on developing a parsimonious program impact theory that represents a common understanding about the expected nature of the program. Once program impact theory is fully developed, program process theory can be unpacked as needed, as stakeholders are presented options for evaluation questions about the need for, or implementation of, the program. This will prevent the "overcomplexity monster" for raising its ugly head once again.

Potential for Substantial Time and Resource Savings

A criticism against program theory-driven evaluation is that it is more time consuming and costly than other evaluation approaches (Scriven, 1998). We have found many situations where this is simply not true. For example, the process of developing program impact theory often reveals that a program is not ready for full-scale evaluation (i.e., evaluative assessment; Wholey, 2003). Time and resources can be saved by redirecting efforts toward program development and/or implementation activities rather than toward summative evaluation, which is certain to reveal null effects. Second, evaluation questions are prioritized in this approach, which helps stakeholders decide how to allocate evaluation resources in a cost-effective manner. We have found that developing program impact theory usually enables stakeholders and evaluators to make informed decisions about evaluation design and methods, often leading to cost-effective evaluation (see Donaldson, 2003).

Program Designs Are Often Improved Prior to Evaluation

The interactive, bidirectional approach to developing program theory often reveals that new programs are poorly conceptualized or at least need some fine-tuning. For example, stakeholders may have designed a program that is not consistent with what is known about changing the behaviors of concern. That is, prior research, well-supported theory, or previous evaluations may suggest that more program development is needed before empirical evaluation methods are employed. In this case, program theory development may result in program improvement before any empirical evaluation work is conducted.

Program Implementation Improvement Without Evaluation

Once program theory is developed, sometimes it becomes apparent that a program is not being implemented well enough to affect key mediators or outcomes conceptualized in the program impact theory. For example, there may be inadequate resources or activities in place to affect target mediators. Again, before employing costly empirical evaluation methods, stakeholders can be given the opportunity to improve the implementation of the program and thus prevent the very common occurrence of discovering no effects through a summative evaluation.

Formative Evaluation Can Lead to Goal Change
Instead of Program Improvement

Although certainly not unique to program theory-driven evaluation, the evaluation of the WHI had a substantial formative evaluation or continuous improvement focus. Formative evaluation findings were used to help stakeholders understand implementation issues and the short-term effects of their programs.

A common expectation of emphasizing formative evaluation in the program theory-driven evaluation approach is that empirical evidence or findings will be used to improve the delivery of the program. However, we learned that these data are also used effectively to adjust program goals and expectations. Once stakeholders see evaluation data, they sometimes decide that negative findings suggest initial goals or expectations are "unrealistic." Rather than continue to fail to meet unrealistic goals, they seek approval to make goals and objectives more reachable, given time and resource constraints. As program theory-driven evaluators, we view this as a healthy process under some conditions (e.g., when goals are indeed out of line), but it can also be used as an excuse for poor performance. Evaluators should carefully document stakeholder reactions to formative evaluation data and be an active voice in the discussions about appropriate responses.

Strategies for Managing Evaluation Anxiety Are Often Required

As described, the program theory-driven evaluation of the WHI required regular interaction between the evaluation team and the stakeholders. The fear of a negative evaluation on a regular basis tended to make some stakeholders very anxious at times (particularly those not confident about their performance). Excessive evaluation anxiety (XEA) can result in problems related to (a) gaining access to critical information, (b) lack of cooperation, (c) compromises to the quality of data collected, (d) undermining the validity of findings, (e) lack of data utilization and program improvement, and (f) a general dissatisfaction with program evaluation and evaluators (Donaldson et al., 2002).

One natural response to XEA by evaluators is to emphasize mere description, as opposed to answering evaluation questions and reaching evaluative conclusions (Scriven, 2003). Although the fear of a negative evaluation is greatly reduced and the evaluation team is likely to be more welcomed and popular with some stakeholders (e.g., program staff), we argue this is an inappropriate response by evaluators to XEA (Donaldson et al., 2002). Instead, we used a number of strategies designed to manage evaluation anxiety. For example, we started with the strategy of training our team to expect and accept evaluation anxiety as a natural component of the interaction with stakeholders. Some of the other strategies we used included (a) legitimating opposition to bad evaluation; (b) determining the program psychologic—what stakeholders are hoping the program will do for them personally; (c) discussing purposes of evaluation in detail to avoid fake evaluations; (d) discussing professional standards for program evaluation; (e) discussing why honesty with the evaluator is not disloyalty to the group; (f) providing role clarification on an ongoing basis; and (g) creating opportunities to role model productive ways to respond to and use evaluation findings (e.g., we encouraged the stakeholders to evaluate the evaluation team formally on a annual basis; see Donaldson et al., 2002, for an in-depth discussion of strategies for managing evaluation anxiety). Strategies for managing evaluation anxiety are often needed while working with stakeholders to develop program impact theory and when conducting rigorous, external, formative and summative evaluations using the program theory-driven approach.

Continuous Improvement Feedback: Don't Forget the Positive Findings

A central goal of the program theory-driven evaluation of the WHI was to facilitate continuous program learning, problem-solving, and program improvement throughout the life of the initiative. The concept of *continuous improvement* has been recognized by many organizations as a means to improve customer satisfaction, enhance product quality, lower operating costs, and to improve organizational effectiveness (Donaldson, 2004). When evaluation is used to improve organizational performance, there are many requirements for it to be effective. Such an evaluation must provide feedback that can be used to (a) improve the design of projects and demonstrations, (b) develop the indicators and performance targets needed to improve effectiveness and responsiveness to challenges, (c) monitor how programs are being implemented for diverse populations across multiple locations, (d) identify and describe key accomplishments, challenges, and lessons learned, and (e) offer recommendations for program improvement.

Core challenges in the evaluation of this initiative were how to produce evaluation reports and develop information systems that were useful to multiple stakeholders. Key issues for evaluators included (a) meeting the diverse information needs of multiple stakeholders, (b) focusing reports on high level issues and challenges, (c) synthesizing tremendous amounts of data from multiple sources, (d) providing diplomatic feedback, and (e) keeping reports to a readable length. In community-based settings, another key challenge to using evaluative data for program improvement purposes concerns stakeholder's general lack of understanding of how evaluation fits with their program goals and is useful for achieving other program goals (e.g., dissemination, policy influence, and the like). In addition, evaluation anxiety over perceived negative consequences that may result from improvement feedback can pose a tremendous barrier to utilization of findings by intended users.

A common mistake evaluators can make here is to focus exclusively on the areas that need improvement. This is a natural mistake because it is often how we define our job and describe the value we add to the social programming endeavor. Of course, the problem is that stakeholders may react very negatively to the process if they do not believe their efforts and successes are fairly acknowledged. Table 12.1 summarizes some observations about delivering continuous improvement feedback.

Building Program and Evaluation Capacity

There is a growing recognition of the need for capacity building in organizations that provide social service programs and interventions. In the United States, most organizations providing these types of services are value-driven organizations that are dependent on external resources for operating and sustaining programs. Because of the complexity of social problems and issues they address, they often face a tremendous scarcity of resources, knowledge, and skills needed to develop and implement effective programs. Those managing and staffing these programs, in turn, have limited time and energy (and experience) to spend on developing and implementing evaluation practices that can be used to monitor and improve their organizations. Therefore, externally funded evaluations that support capacity-development efforts arguably represent one of the most valuable resources these organizations have for building program capacity.

Unfortunately, a lack of understanding and/or appreciation of evaluation pose key barriers to prioritizing and allocating precious resources toward evaluation. For example, it is not uncommon for organizational members to question the credibility of evaluators, balk at the costs, and highlight potential undesirable consequences of using external evaluation (Donaldson, 2001b). This becomes particularly salient when the demands of evaluation compete with resources needed to meet program service goals.

One of the key dilemmas to building capacity in the community-based and human service organizations participating in the WHI was the loss of capacity due to relatively frequent turnover in program personnel. With respect to building program evaluation capacity, the consequences of personnel turnover include the loss of grantee knowledge and understanding of evaluation goals, terminology, and approach. In addition, turnover contributes to a loss of ownership and buy-in to the evaluation design and methodology among newcomers. In some cases, staff turnover may result in evaluation measures, tools, and resources being taken by departing individuals, leaving newcomers with little to no evidence of evaluation progress that was made. Similar losses are

TABLE 12.1

Summary of Observations About Delivering Continuous Improvement Feedback

---

- Establishing a safe environment that facilitates open communication and a willingness to share both positive and negative information is essential for facilitating utilization of continuous improvement feedback. To achieve this, the evaluator must establish trust, rapport, and credibility with program stakeholders, and oftentimes, engage in long-term relationship management.
- It is not easy overcoming negative perceptions about evaluation and expectations that evaluation plays a punitive role in community-based organizations. Achieving this requires careful listening to stakeholder concerns, values, and needs, and dedicating time for discussing and prioritizing the information needs of multiple stakeholders. Related to this, it is important that evaluators possess awareness and sensitivity to the cultural values of intended users of improvement-oriented feedback.
- Continuous improvement feedback is useful only to the extent that it is timely and meets the core information needs of program stakeholders. This entails providing frequent reports and feedback. Evaluators delivering this type of feedback should allocate sufficient staff and resources toward writing numerous evaluation reports. Because information needs will vary for different program stakeholders, and will change over time, it is important that feedback reports be tailored to different target audiences.
- It is helpful to revisit information needs and adjust data collection practices accordingly as programs evolve and grow.
- In a continuous-improvement model, evaluators should not underestimate the important role of discussing the evaluation findings with program stakeholders. Evaluation reports cannot substitute for dialogue when it comes to understanding the complex issues facing human service organizations. Feedback reports are critical for examining facts about program achievements and challenges. However, it is only through ongoing dialogue about the meaning and implications of those facts that program stakeholders can identify lessons learned and use evaluation findings for problem-solving and decision making.
- Balancing program improvement feedback with feedback pertaining to program accomplishments and strengths is essential for reducing program stakeholders' resistance to discussing and using evaluation findings. Failure to acknowledge the context in which the findings take place, including efforts to address program issues and key accomplishments will result in defensiveness toward findings and a lack of utilization of findings by intended users.

---

experienced with respect to programmatic knowledge and skills that have been learned by staff over time. Unfortunately, there remains a lack of clear guidelines and practices for building and sustaining capacity in human service organizations. We have presented some of the lessons learned in this evaluation related to building program and evaluation capacity in Table 12.2.

TABLE 12.2

Lessons Learned Related to Building Program and Evaluation Capacity

---

- Evaluators must expect and plan for staff turnover in community-based and human service programs. This entails initial and ongoing instruction and training in evaluation terms and procedures for new staff and program directors.
- Program stakeholders may not see the connection between program evaluation and the uses of findings for facilitating program improvements. To be effective, stakeholders must understand how the evaluation fits with their overall program goals and how they will benefit from incorporating evaluation practice in their organization. Evaluators should expect that users will hold negative stereotypes about evaluation's punitive role and must work to clarify the many uses of evaluation for intended users.
- Evaluators should plan on developing clear and straightforward evaluation tools and resources for program stakeholders. Key resources may include evaluation plans, evaluation procedure manuals, glossary of evaluation terms, evaluation roadmaps, evaluation training, evaluation training manuals, and frequent contact and communications with program stakeholders.
- Evaluators should not assume that technical assistance providers and program management teams understand the role and practice of evaluation. Lack of understanding, commitment, and support from these groups can derail evaluation efforts in human service organizations. To minimize role conflict and to prevent miscommunications about evaluation practice, evaluators need to build evaluation knowledge and skills in these groups as well. It is also critical that they buy in to the evaluation and support evaluation efforts.
- Evaluators must be strategic and sensitive in asking for time and involvement from busy people in evaluation efforts, and they must be sure they are interacting with the right people around relevant issues. Evaluation capacity cannot be increased in community-based and human service organizations without addressing the practical implications of conducting evaluations in these types of organizations. Data collection and tracking efforts that are overly cumbersome will not be supported.
- Evaluators will typically have to work to build and sustain evaluation capacity. Building effective relationships with intended users requires an understanding and sensitivity to the populations served, the sociocultural and economic contexts in which the program operates, and the values and needs of program constituents. It also requires ongoing relationship building and training in evaluation procedures.
- Evaluators need strong interpersonal skills in building relationships, facilitating groups, and managing conflict. They must communicate effectively and be keenly sensitive to human and organizational behavior issues.

Managing the Formative to Summative Evaluation Transition

Using evaluation results to improve a program is fundamentally different from rendering judgment about merit, worth, or significance for an external audience. In the formative phase of the program theory-driven evaluation of the WHI, questions tended to be more open-ended and focused on identifying program strengths and weaknesses. In the summative phase, in contrast, we focused on using explicit criteria of effectiveness in forming the basis for evaluative conclusions. A key challenge to these contrasting purposes of evaluation is that they often come into conflict because the information needed for program management and improvement is different from the data needed for accountability. These approaches, in turn, lead to different experiences for program managers and staff. Whereas conducting formative evaluation is experienced as developmental, conducting cumulative or summative evaluation is sometimes perceived to be judgmental and punitive.

Credible Evidence: Often the Achilles Heel of Effectiveness Evaluation

Step 3 of the program theory-driven evaluation process involves exploring ways to gather credible evidence to answer the evaluation questions of interest. Most effectiveness evaluation contexts require this discussion to take place in light of time, resource, and feasibility constraints. That is, it is rare to be in a position to have the time and resources to design an elaborate multimethod design that rules out all threats to validity and provides the highest level of confidence in the answers to each key evaluation question. It should also be noted that in the WHI evaluations and in effectiveness evaluations more generally, there seems to be few examples of stakeholders committing scarce resources to pursuing credible evidence to uncover harmful side effects (even though this is a very critical component of sound evaluation). It seems more common today that evaluators and stakeholders satisfy rather than optimize, when selecting methods to gather data to answer key evaluation questions. Further complicating the quest for credible evidence is the lack of agreement in the field about the strengths and weaknesses of a variety of evaluation methods for gathering credible evidence (see American Evaluation Association, 2004; Donaldson & Christie, 2005; Donaldson, Christie, & Mark, in press).

We have observed in effectiveness evaluation practice that it is sometimes difficult to reach consensus on evaluation methods. Sometimes this results from stakeholder disagreements about what constitutes credible evidence, whereas at other times this can be due to stakeholders not willing to commit resources (or they do not have access to resources) to support a strong evaluation design and data collection effort. A curious observation is that the evidence used to answer evaluation questions seems to be most often challenged when the answer to an evaluation question is undesirable to the challenger. That is, the same methods and evidence may be praised, or at least remain unchallenged, in light of desirable answers to key evaluation questions. The main strategy I suggest here is to attempt to reach agreement about design and methods before collecting data and to be prepared to address challenges about the quality of the evidence collected, especially when null or negative findings emerge.

It is now common, for theorists at least, to recommend that evaluations themselves should be evaluated (Scriven, 2003). The evaluation of an evaluation is often referred to as a *metaevaluation*. That is, in an ideal evaluation world, every evaluation would have the benefit of an objective, external evaluation that reveals the strengths and limitations of the evaluation itself. A metaevaluation can be used to render an independent judgment about the credibility of the evidence that was used to reach evaluative conclusions. Although this appears to be a desirable idea in theory, it is often difficult to convince stakeholders in effectiveness evaluation to commit additional resources for this activity. However, in some situations, metaevaluation is a technique that can be used to assess and bolster the credibility of evidence used to answer evaluation questions.

REFLECTIONS ON THE BUNCHE–DA VINCI PROPOSAL

The WHI applications of program theory-driven evaluation science give us a window into the challenges that practitioners may encounter when they use program theory-driven evaluation science to provide evaluation services in dynamic "real world" settings. Alkin and Christie (2005) used a different approach, a case scenario exercise, to attempt to reveal similar information about contemporary evaluation practice. Specifically, the exercise was designed to illustrate how evaluators who use different theoretical approaches would propose to evaluate the exact same program, at the exact same stage in its history. That is, by holding the evaluand or case constant (which can only be done well using a hypothetical case exercise), evaluation proposals based on different theoretical perspectives were compared and contrasted to develop a better understanding of the relationship between theory and practice (Alkin & Christie, 2005).

This exercise resulted in four detailed proposals for how evaluation services could be provided. The case (chap. 11) and a detailed proposal using program theory-driven evaluation science (chap. 12) were provided in this volume to illustrate how program theory-driven evaluation science could be deployed to help the Bunche–Da Vinci Partnership Academy Program better meet the needs of a diverse and disadvantaged group of elementary students.

What can we learn about program theory-driven evaluation science from this case exercise? First, the proposal shows how to use this approach to develop a realistic and fair contract to provide evaluation services. An interactive discussion with stakeholders over several meetings attempted to simulate the type of interactions and questions that can typically arise in evaluation practice. Next, a rather detailed plan is presented to show how the three steps of the program theory-driven evaluation process could be carried out to provide practical and cost-effective external evaluation services in this case. This information may be useful to evaluation practitioners required to describe how to use program theory-driven evaluation science to address a new program or problem and it provides a model of how to structure a contract for evaluation services using this approach.

Other major benefits of the Bunche–Da Vinci exercise are (a) to learn how this theoretical perspective, program theory-driven evaluation science is translated into the mechanics of basic evaluation practice; and (b) how this translation compares with three other practical translations of evaluation theories (value-engaged, social betterment, and building evaluation capacity approaches). Alkin and Christie (2005) provided a comparative analysis of the four vastly different evaluation proposals. They compared and contrasted the proposals across the themes of depth of participation by stakeholders, concerns for evaluation use, use of program theory, attention to social justice, and the evaluation methods that were used. Each theorist was given the opportunity to respond to their analysis, and their responses provided even more insight into how to apply their approach in practice. The comparative analysis provided by Alkin and Christie (2005), as well as the other three proposals (Greene, 2005; Henry, 2005; King, 2005), are likely to provide an additional understanding of how to apply program theory-driven evaluation science.

LINKS TO EFFICACY EVALUATIONS

Lipsey and Wilson (1993) metaanalyzed the findings of more than 10,000 evaluations of psychological, educational, and behavioral programs. They concluded that most of these efficacy evaluations were based on only crude "black box" outcome evaluations, paying little attention to program impact theory or potential mediators or moderators. It was also suggested that the proper agenda for the field of evaluation to move forward was to begin to focus evaluations on (a) which program components are most effective, (b) the mediating causal processes through which programs work, and (c) on the characteristics of the participants, service providers, and contexts that moderate the relationships between a program and its outcomes.

Serious dialogue and applications of program theory-driven evaluation are now prevalent throughout most regions across the global evaluation landscape and within most of the major evaluation associations. For example, recent volumes, papers, and visions for future applications of theory-driven evaluation have been discussed at length within the European evaluation community (e.g., Pawson & Tilley, 1997; Stame, 2004; Stern, 2004; van der Knapp, 2004). Published examples of theory-driven evaluations now exist from many parts of the world, including but not limited to Tiawan (Chen, Wang, & Lin, 1997); Germany (Bamberg & Schmidt, 1997); Britian (Tilley, 2004); South Africa (Mouton & Wildschut, 2006); Canada (Mercier, Piat, Peladeau, & Dagenais, 2000); as well as the United States (Bickman, 1996; Cook, Murphy, & Hunt, 2000; Donaldson, Graham, & Hansen, 1994; Donaldson, Graham, Piccinin, & Hansen, 1995; Reynolds, 1998; West & Aiken, 1997). However, the vast majority of these writings and applications deal mostly with efficacy evaluation, often illustrating the power of experimental and quasi-experimental designs to determine program impact and to identify generalizeable mechanisms that can be used again to prevent and ameliorate problems elsewhere. Most of these applications demonstrate the value of collecting and analyzing large data sets and using some of the most sophisticated analytic methods available, such as latent structural and hierarchal linear modeling (see Shadish et al., 2002; Reynolds, 1998).

This volume is intended to demonstrate that many of the same principles that apply to using program theory-driven efficacy evaluation extend with some modification to more common effectiveness evaluation problems. I have shown how program theory-driven evaluation science can be used when stakeholders do not have the luxury of the resources or the ability to design large-scale, randomized controlled trials, and the like, that typically extend over long periods of time. That is, program theory-driven evaluation science can be easily used (a) when stakeholders are required or desire to be included in decisions about the evaluation process and design, (b) when timely and cost-effective formative and summative evaluation is required, and (c) when some or many of the common methods discussed in most evaluation texts today are not feasible.

I do not mean to imply that effectiveness and efficacy evaluation are equivalent—they are not. Rather, I am trying to emphasize here that program theory-driven evaluation science is robust enough to handle both types of evaluation, perhaps better than most other evaluation approaches, and is not dependent on method constraints and proclivities. Both efficacy and effectiveness program theory-driven evaluations have important roles to play in helping to build an evidence base to shed light on how to prevent and solve the social, educational, health, community, and organizational problems that are likely to confront us in the years ahead.

CONCLUSION

This chapter concludes with observations and analyses about the use of program theory-driven evaluation science in contemporary evaluation practice. Insights and lessons learned from the dynamic interplay between evaluators and stakeholders in effectiveness evaluation practice, as well as from a case exercise resulting in a specific evaluation proposal, have been discussed in an effort to improve future applications of program theory-driven evaluation science. It has been noted often in recent times that (a) practical advice, (b) explicit examples of evaluation theory being implemented in practice, (c) written insights and lessons learned from applications of evaluation in "real world" settings, and (d) descriptive data from evaluation practice are sorely needed to advance our understanding of the new realities of the growing profession and discipline of evaluation science. It is my hope that the applications and observations presented here at least inch us in the right direction toward filling this gap.

13

Practical Implications of the Emerging Transdiscipline of Evaluation Science

I have explored the use of program theory-driven evaluation science to improve lives of hundreds of thousands of people living in the state of California. Systematic data were collected, analyzed, disseminated, and used for decision making to support the development of effective programs to improve the productivity and well-being of California workers and their families. These applications of program theory-driven evaluation science were presented in detail to help advance the understanding of the challenges and benefits of using this approach in practice, as well as to provide insights into how to improve the use of evaluation science to develop people, programs, and organizations.

In this chapter, I attempt to place program theory-driven evaluation science in a broader context. Specifically, I discuss the practical implications of evaluation science as an emerging transdiscipline. This discussion includes views on how evaluation practitioners might take advantage of recent advances in knowledge about evaluation theory and practice. Finally, I end by suggesting where I think program theory-driven evaluation science fits within this emerging transdiscipline and present some new directions for improving its ability to enhance effectiveness and well-being.

EVALUATION SCIENCE: AN EMERGING TRANSDISCIPLINE

Even a decade ago, it would have been difficult to imagine how pervasive, productive, and global in scope the field of evaluation would become. The zeitgeist of accountability and evidence-based practice is

237

now widespread across the professions and across the world. Fields as wide-ranging as medicine, public health, education, mental health, social work, political science, international development, economics, psychology, organizational behavior, and management have bought into the notion that advancing an evidence-based practice is critical to achieving a bright future for their field. Consequently, organizations from a wide range of sectors such as health care, education, business, the nonprofit and social sectors, philanthropy, and federal, state, and local governments are being asked to evaluate their practices, programs, and policies at an increasing rate in an effort to promote human welfare and achievement (Donaldson & Scriven, 2003b).

A key indicator of the emergence and globalization of evaluation is the dramatic increase in the number of professionals who have become members of organized evaluation associations and societies. In 1990, there were approximately five major professional evaluation associations, whereas today there are more than 50 worldwide (Mertens, 2003, 2005; Russon, 2004). For example, Table 13.1 lists a sample of organizations that illustrate this new professional networking and development activity. An international alliance has been formed to link these professional organizations in an effort to share knowledge about how to improve the practice of evaluating a wide range of programs, policies, projects, communities, organizations, products, and personnel, and to promote social betterment worldwide (Russon, 2004).

These new realities of evaluation science demonstrate its practical quality of providing tools for and enhancing other disciplines, as well as improving the effectiveness of professionals from across a wide variety of fields (Donaldson & Christie, 2006). These are some of the characteristics of what Scriven (1991) described as "evaluation as a transdiscipline." In later work, Scriven (2003) shared his vision of a positive future for evaluation science:

> I hope and expect that the essential nature of evaluation itself will crystallize in our minds into a clear and essentially universal recognition of it as a discipline, a discipline with a clear definition, subject matter, logical structure, and multiple fields of application. In particular, it will, I think, become recognized as one of the elite group of disciplines, which I call transdisciplines. These disciplines are notable because they supply essential tools for other disciplines, while retaining an autonomous structure and research effort of their own. (p. 19)

Scriven's vision for what evaluation science might become appears more realistic as each year passes. There is now plenty of evidence that evaluation science is used by various leaders, managers, administrators, educators, policymakers, researchers, philanthropists, service providers, and the like, to:

TABLE 13.1

Sample of Professional Evaluation Organizations

- African Evaluation Association
- American Evaluation Association
- *Association Comorienne de Suivi et Evaluation*
- *Associazione Italiana de Valuatazione*
- Australasian Evaluation Society
- Bangladesh Evaluation Forum
- Botswana Evaluation Association
- Brazilian M&E Network
- Burundi Evaluation Network
- Canadian Evaluation Society
- Central American Evaluation Association
- Danish Evaluation Society
- *Deutsche Gesellschaft für Evaluation*
- Egyptian Evaluation Association
- Eritrea Evaluation Network
- Ethiopian Evaluation Association
- European Evaluation Society
- Finnish Evaluation Society
- Ghana Evaluators Association
- Ghana Evaluation Network
- International Program Evaluation Network (Russia/NIS)
- Israeli Association for Program Evaluation
- Japanese Evaluation Association
- Kenya Evaluation Association
- Korean Evaluation Association
- *La Societe Francaise de l'Evaluation*
- Malawi M&E Network
- Malaysian Evaluation Society
- Namibia Monitoring Evaluation and Research Network
- Nepal M&E Forum
- Nigerian Evaluation Association
- Programme for Strengthening the Regional Capacity for Evaluation of Rural Poverty Alleviation Projects in Latin America and the Caribbean (PREVAL)
- *Reseau Malgache de Suivi et Evaluation*
- *Reseau Nigerien de Suivi et Evaluation*
- *Reseau Ruandais de Suivi et Evaluation*
- *Societe Quebecoise d'Evaluation de Programme*
- *Societe Wallonne de l'Evaluation et de la Prospective*
- South African Evaluation Network
- Spanish Public Policy Evaluation Society
- Sri Lanka Evaluation Association
- Swiss Evaluation Society
- Thailand Evaluation Network
- Ugandan Evaluation Association
- UK Evaluation Society
- *Utvarderarna (Sweden)*
- Zambia Evaluation Association
- Zimbabwe Evaluation Society

- Make better decisions.
- Produce practical knowledge.
- Promote organizational learning.
- Improve practice, programs, policies and organizations.
- Determine merit, worth, and significance.
- Comply with accountability demands.

Donaldson and Christie (2006) pointed out that evaluation science is making its way into the college and university curriculum in numerous fields, and can now be thought of as a powerful enhancement of professional training in a wide variety of areas.

At the same time evaluation science is being used to enhance other disciplines and develop cumulative knowledge about interventions designed to prevent and solve a wide range of contemporary problems in our global society, it appears to also be developing a unique knowledge base and research effort of its own. New directions and practical implications of recent advances of knowledge about how best to practice evaluation science is now discussed.

Using Evaluation Theory in Practice

In his presidential address for the American Evaluation Association, Shadish (1998) asserted:

> Evaluation theory is who we are, all evaluators should know evaluation theory because it is central to our professional identity. It is what we talk about more than anything else, it seems to give rise to our most trenchant debates, it gives us the language for talking to ourselves and others, and perhaps most important, it is what makes us different from other professions. Especially in the latter regards, it is in our own self-interest to be explicit about this message, and to make evaluation theory the very core of our identity. Every profession needs a unique knowledge base. For us, evaluation theory is that knowledge base. (p. 1)

As discussed in chapter 1, evaluation theory prescribes how we should practice evaluation science. Alkin (2004) suggested that, in the context of the current stage of the evaluation profession's maturity, evaluation theories may be more accurately described as "approaches and models" that serve as exemplars of how best to practice evaluation. They are largely a prescriptive set of rules, prohibitions, and guiding frameworks that specify what a good or proper evaluation is. They describe what should be, rather than what is.

I have also discussed in this volume the very large collection of evaluation theories, approaches, models, or brands of evaluation. Some of these theories are similar and/or complementary, others fundamentally disagree about basic assumptions and principles related to how best to practice contemporary evaluation science. Although great strides have been made in recent years to understand the differences, strengths, and weaknesses of modern theories of evaluation practice (e.g., Alkin, 2004; Alkin & Christie, 2005; Donaldson & Scriven, 2003b), there is currently no agreed-on best "evaluation theory," integrative framework of evaluation theories, or best way to practice evaluation.

The practical implications of this current "state of affairs" for practicing evaluators are vitally important. First, evaluation practitioners will be better equipped to face the challenges of evaluation practice if they are aware of the history and diversity of theories of evaluation practice. For example, some clients might request a particular approach or brand of evaluation science. Practitioners need to be prepared to discuss the strengths and weaknesses of that approach in relation to the evaluation assignment, and perhaps suggest other options that could be more appropriate or effective for the specific evaluation problem at hand. Second, it is not uncommon that a practitioner's work is criticized or metaevaluated negatively by someone using a different set of assumptions about what constitutes good evaluation practice. Knowledge about evaluation theory can help sort out differences of professional judgment and opinion.

As we have discussed, the transdiscipline of evaluation science appears to be growing rapidly in new directions. There is a now a wide range of purposes for conducting evaluations. Purposes such as program and organizational improvement, oversight and compliance, assessment of merit and worth, and knowledge development (Mark et al., 2000) seem to be sanctioned by most modern theories of evaluation practice. Whereas other purposes such as using evaluation data to support a political position, to enhance public relations, to support a predetermined decision of any sort, or to sell products may be more suspect depending on how well the evaluation design conforms with professional evaluation theory and standards of accuracy and objectivity.

There is also a widening range of applications for the transdiscipline of evaluation science. Although most of the early writing on evaluation science seemed heavily rooted in program and policy evaluation, members of the evaluation profession are extending practice well beyond these two common types of evaluations. For example, the American Evaluation Association now describes its membership in the following light:

> Evaluation is a profession composed of persons with varying interests, potentially encompassing but not limited to the evaluation of programs, products, personnel, policy, performance, proposals, technology, research, theory, and even of evaluation itself. (American Evaluation Association, 2004)

Theories of evaluation practice are now challenged to specify how far they extend across the various types of evaluations. Of course, some evaluation theories might specifically focus on one type of evaluation (e.g., program evaluation), whereas others may apply more broadly. Modern practitioners need to be aware of their scope of practice, how the evaluation theory or model they follow compares and contrasts with others, and how it applies or works across the different applications of evaluation science. The awareness of the diversity of evaluation theory and growing number of applications of evaluation science alone promises to help those who will practice evaluation in the future to be more informed and effective. It is no longer acceptable to practice evaluation using only methods of the basic social sciences, nor to teach practitioners only one evaluation approach or evaluation theory from which to operate and evaluate.

Evaluation practitioners in the emerging transdiscipline must also be aware of a much broader range of evaluation designs, tools, and data collection methods. In addition, they must be equipped to deal with many types of diversity and cross-cultural issues including how best to provide culturally competent evaluation services (see Thompson-Robinson, Hopson, & SenGupta, 2004). The debates about whether it is more desirable to use quantitative or qualitative methods for evaluation have been replaced by the search for guidelines that specify the strengths and weaknesses of the use of a specific method for a specific evaluation task (e.g., needs assessment, evaluative assessment, program implementation analysis, expressing and assessing program theory, process evaluation, impact assessment, personnel evaluation, cost benefit analysis, cost effectiveness analysis, and the like) and for specific evaluation purposes (e.g., to improve programs, organizational learning, oversight and compliance, determining merit and worth, selection and compensation, policy analysis, a wide range of decision making purposes, and so on; Donaldson, Christie, & Mark, in press). Each method choice that evaluators make should be viewed in terms of how well the evaluation design and methods produce data and information that meet the evaluation standards of accuracy, feasibility, propriety, and use (Joint Committee on Standards for Education Evaluation, 1994). Practicing evaluators are often required to convince stakeholders and decision makers that evaluation findings and conclusions are based on credible evidence. That is, a thoughtful and critical analysis of the plausibility of threats to validity and potential bias in evaluation data used to reach evaluative conclusions should be considered standard practice in all types of evaluations across the transdiscipline.

Developing an Evidence-Based Evaluation Practice

Most theories of evaluation practice are based more on philosophy and experience than on systematic evidence of their effectiveness. Although there is often discussion at professional gatherings and, to some extent, in the evaluation literature for the need to accumulate research results to inform evaluation practice, evaluation theories of practice remain largely prescriptive and unverified. This presents a challenge, because, one of the defining features of a sound transdiscipline is that it is able to develop a unique evidence base or research effort of its own, independent of its application areas (Scriven, 2003). Another practical implication of the emerging transdiscipline is the expanded need for research across the application areas, and the growing menu of topics that could be pursued by evaluation scholars interested in determining how best to practice evaluation science.

Mark (2003) expressed serious concern about the lack of evidence-based research on evaluation. He would like to see work developed that focuses on:

- Purposively and prospectively setting up alternative methods to compare.
- Studies that systematically compare findings, retrospectively, across different types of evaluation.
- Studies that track evaluation use as a function of the evaluation approach.

Besides conducting evaluations, the second preoccupation of evaluators is theoretical exposition and advocacy which covers the pages of most books and articles on evaluation. Although Mark (2003) acknowledges a place for prescriptive theory and advocacy, he asserts that, without a sound evidence base for theory development, the evaluation profession will not be able to significantly improve its evaluation theory and practice.

Henry and Mark (2003) presented an ambitious agenda for research on evaluation that embraced the following themes:

- Research on evaluation outcomes.
- Comparative research on evaluation practice.
- Metaevaluation.
- Analog studies.
- Practice component studies.
- Evaluation of technical assistance and training.

Systematic evidence to inform the areas just mentioned promises to resolve some of the debates and disputes about how best to practice evaluation and could dramatically enhance the effectiveness of practitioners faced with the diversity of applications, approaches, and methods presented by the changing nature of evaluation science.

Although there is much ground left to plough and many new topics to investigate in the emerging transdiscipline, we have witnessed some important scholarly contributions in recent years. For example, there have been significant advances in understanding the range of approaches and theories for conducting evaluations (Alkin, 2004; Donaldson & Scriven, 2003); how closely evaluation theory reflects actual practice (Alkin & Christie, 2005; Christie 2003; Fitzpatrick, 2004); research about the best ways to ensure the productive use of evaluation findings (Henry & Mark, 2003); strategies for overcoming excessive evaluation anxiety (Donaldson et al., 2002); improving the relationships between evaluators and stakeholders (Donaldson, 2001b); and the development of evaluation competencies (Ghere, King, Stevahn, & Minnema, 2006); standards of practice (e.g., Joint Committee on Standards for Education Evaluation, 1994); and guiding principles (e.g., American Evaluation Association, 2004). This expanding knowledge base about evaluation practice itself is being used to inform and improve the growing number of applications of evaluation science.

How will the evaluation practitioner possibly stay abreast of the latest advances in evaluation science, particularly evidence-based developments? After all, House (2003) noted that many practitioners are part-time evaluators who are not trained or well-versed in evaluation theory. Further, there is believed to be a high turnover of practitioners who enter and exit evaluation practice every year. House observed that this situation minimizes the incentives for individual practitioners to invest time and effort in learning sophisticated theories.

The part-time nature of evaluators, high turnover, the lack of professional training by many required to conduct evaluations, and the rapidly changing nature of evaluation practice has led to a growing demand for evaluation education and training. The strong need for training seems to be spawning a new cadre of participation-oriented (versus competency-based) professional development training institutes in evaluation. Unfortunately, at the same time the need for competency-based evaluation education and training is at its highest point in some time, the number of rigorous university-based evaluation programs appears to be declining (Engle & Altschuld, 2004). This trend has serious implications for the quality of work in the emerging transdiscipline. More support for and attention to building a broad network of university-based programs that provide competency-based degrees is critical to preparing a new generation of evaluators equipped to take on the challenges of the emerging transdiscipline of evaluation science.

IMPLICATIONS FOR THE FUTURE OF PROGRAM THEORY-DRIVEN EVALUATION SCIENCE

How will program theory-driven evaluation science fare in this broader context of the emerging transdiscipline of evaluation science? Program theory-driven evaluation science strives to be an evolving and integrative theory of program evaluation practice. As was introduced in chapter 1, Shadish et al. (1991) concluded that theory-driven evaluation was initially a very ambitious attempt to bring coherence to a field in considerable turmoil and debate, and the integration was more or less successful from topic to topic. Continuing the strategy of incorporating new knowledge, techniques, and methods into an evolving evaluation theory of program theory-driven evaluation science seems essential for it to remain productive in the emerging transdiscipline.

Within this broader context of evaluation science, program theory-driven evaluation specifies feasible practices for program evaluators. I am not aware of attempts to extend its application to the evaluation of personnel, products, policies, projects, performance, proposals, technology, or research. One question that needs to be asked moving forward is: Should program theory-driven evaluation science be modified and extended to address additional applications of evaluation science? Or, is it more appropriate to remain focused on developing a deeper knowledge base about how best to evaluate programs that attempt to develop people, programs, and organizations?

Program theory-driven evaluation science is currently a rather robust approach to program evaluation across disciplines and professions. For example, program theory-driven evaluation has been used to improve and determine the merits of programs in public health, education, mental health, child development, applied psychology, career and organizational development, and the like. The notion of using a conceptual framework or program theory grounded in relevant substantive knowledge to guide evaluation efforts seems consistent, or at least compatible, with the values of those working to advance these other disciplines. However, the effectiveness of program theory-driven evaluation science could be greatly improved in the years ahead if a sound evidence base is developed to better inform practice. Some of the questions regarding this research agenda might include how best to:

- Engage stakeholders in the three-step process.
- Develop program process and program impact theories.
- Prioritize evaluation questions.
- Select methods to answer evaluation questions.

- Control bias in both efficacy and effectiveness evaluations.
- Produce sound evaluation conclusions and recommendations.
- Ensure evaluation use and disseminate lessons learned.
- Meet evaluation standards.
- Provide cost-effective evaluation services using this approach.
- Educate and train practitioners to use program theory-driven evaluation science.

Although a research agenda of this magnitude may seem like a tall order given the current research capacity of the transdiscipline, strides toward building this type of evidence-based foundation could pay large dividends to the field down the road. For example, if the evidence supports using and/or improving this particular theory of practice, this approach should be given a large role in the emerging transdiscipline of evaluation science. If it turns out research evidence uncovers a better alternative or alternatives to program theory-driven evaluation science, or that it is not an effective way to practice evaluation, new directions should be pursued in order to advance evaluation science.

CONCLUSION

This chapter aimed to describe the emerging transdiscipline of evaluation science and to show how program theory-driven evaluation science might fit into this broader context. Program theory-driven evaluation science today is a theory of practice focused on program evaluation. Its applications extend across a wide range of professions and disciplines. Efforts to move it from a prescriptive theory of practice toward a theory of practice based on research evidence seem to hold great promise for advancing evaluation science. If sound research supports that it is an effective way to provide evaluation science services, some of its principles and practices may be exported to inform other applications of evaluation science.

This book has now explored in depth strategies and applications of program theory-driven evaluation science. These strategies were first described and then illustrated across a number of "real world" applications. The ultimate purpose of trying to better understand and refine program theory-driven evaluation science for contemporary practice is to help improve this tool for creating a science of developing people, programs, and organizations.

Although it remains to be seen whether evaluation science will realize its potential in the years ahead, it is my hope that the strategies, observations, and lessons learned presented in this volume will contribute to the understanding of how to address our most pressing human problems as the 21st century unfolds.

REFERENCES

Adkins, J. A. (1999). Promoting organizational health: The evolving practice of occupational health psychology. *Professional Psychology, 30*(2), 129–137.

Adler, N. E., Marmot, M., McEwen, B. S., & Stewart, J. (1999). *Socioeconomic status and health in industrial nations.* New York: New York Academic Sciences.

Aley, J. (1995). Where the jobs are. *Fortune, 132,* 53–56.

Alkin, M. C. (Ed.) (2004). *Evaluation roots.* Thousand Oaks, CA: Sage.

Alkin, M. C., & Christie, C. A. (2004). An evaluation theory tree. In M. C. Alkin (Ed.), *Evaluation roots* (pp. 12–65). Thousand Oaks, CA: Sage.

Alkin, M. C., & Christie, C. A. (Eds.) (2005). *Theorists' models in action.* San Francisco, CA: Jossey-Bass.

American Evaluation Association. (2004, July). *Guiding principles for evaluators.* Retrieved February 24, 2004, from http://www.eval.org

Bamberg, S., & Schmidt, P. (1997). Theory driven evaluation of an environmental policy measure: Using the theory of planned behavior. *Zeitschrift für Sozialpsychologie, 28*(4), 280–297.

Baron, R. M., & Kenny, D. A. (1986). The moderator–mediator variable distinction in social psychological research: Conceptual, strategic, and statistical considerations. *Journal of Personality and Social Psychology, 51,* 1173–1182.

Bickman, L. (1987). The functions of program theory. *New Directions for Program Evaluation, 33,* 5–18.

Bickman, L. (1996). A continuum of care: More is not always better. *American Psychologist, 51*(7), 689–701.

Brousseau, R., & Yen, I. (2000). *Reflections on connections between work and health.* Thousand Oaks, CA: California Wellness Foundation.

Bureau of Labor Statistics. (1999, July). Unemployed persons by state, and duration of unemployment. Retrieved July 1, 1999, from http://www.bls.gov/schedule/archives/empsit_nr.htm#1999

The California Work and Health Survey (1998, September). San Francisco, CA: University of California.

Campbell, B., & Mark, M. M. (2006). Toward more effective stakeholder dialogue: Applying theories of negotiation to policy and program evaluation. *Journal of Applied Social Psychology, 36*(12), 2834–2863.

Centers for Disease Control. (1999) *Centers for Disease Control Program Evaluation Framework (1999).* 48 (RR11), 1–40.

Chen, H. T. (1990). *Theory-driven evaluations.* Newbury Park, CA: Sage.

Chen, H. T. (1997). Applying mixed methods under the framework of theory-driven evaluations. *New Directions for Evaluation, 74,* 61–72.

Chen, H. T. (2004). The roots of theory-driven evaluation: Current views and origins. In M. C. Alkin (Ed.), *Evaluation roots* (pp. 132–152). Thousand Oaks, CA: Sage.

Chen, H. T. (2005). *Practical program evaluation: Assessing and improving planning, implementation, and effectiveness.* Newbury Park, CA: Sage.

Chen, H. T., & Rossi, P. H. (1983). Evaluating with sense: The theory-driven approach. *Evaluation Review, 7,* 283–302.

Chen, H. T., & Rossi, P. H. (1987). The theory-driven approach to validity. *Evaluation & Program Planning, 10,* 95–103.

Chen H. T., Wang J. C. S., & Lin, L. H. (1997). Evaluating the process and outcome of a garbage reduction program in Taiwan. *Evaluation Review, 21*(1), 27–42.

Christie, C. A. (Ed.). (2003). *The practice-theory relationship in evaluation.* San Francisco, CA: Jossey-Bass.

Cook, T. D. (1985). Post-positivist critical multiplism. In R. L. Shotland & M. M. Mark (Eds.), *Social science and social policy* (pp. 21–61). Beverly Hills, CA: Sage.

Cook, T. D., Murphy, R. F. & Hunt, H. D. (2000). Comer's school development program in Chicago: A theory-based evaluation. *American Educational Research Journal, 37*(2), 535–597.

Cousins, J. B., & Leithwood, K. A. (1986). Current empirical research on evaluation utilization. *Review of Educational Research, 56*(3), 331–364.

Crano, W. D. (2003). Theory-driven evaluation and construct validity. In S. I. Donaldson & M. Scriven (Eds.), *Evaluating social programs and problems: Visions for the new millennium* (pp. 111–142). Mahwah, NJ: Lawrence Erlbaum Associates.

Cummings, T. G., & Worley, C. W. (2005). *Organization development and change* (8th ed.). Cincinnati, OH: South-Western College Publishing.

Davidson, E. J. (2005). *Evaluation methodology basics: The nuts and bolts of sound evaluation.* Thousand Oaks, CA: Sage.

Donaldson, S. I. (1995a). Worksite health promotion: A theory-driven, empirically based perspective. In L. R. Murphy, J. J. Hurrel, S. L. Sauter, & G. P. Keita (Eds.), *Job stress interventions* (pp. 73–90). Washington, DC: American Psychological Association.

Donaldson, S. I. (1995b). Peer influence on adolescent drug use: A perspective from the trenches of experimental evaluation research. *American Psychologist, 50,* 801–802.

Donaldson, S. I. (2001a). Mediator and moderator analysis in program development. In S. Sussman (Ed.), *Handbook of program development for health behavior research* (pp. 470–496). Newbury Park, CA: Sage.

Donaldson, S. I. (2001b). Overcoming our negative reputation: Evaluation becomes known as a helping profession. *American Journal of Evaluation, 22(3),* 355–361.

Donaldson, S. I. (2003). Theory-driven program evaluation in the new millennium. In S. I. Donaldson & M. Scriven (Eds.), *Evaluating social programs and problems: Visions for the new millennium* (pp. 111–142). Mahwah, NJ: Lawrence Erlbaum Associates.

Donaldson, S. I. (2004). Using professional evaluation to improve the effectiveness of nonprofit organizations. In R. E. Riggio & S. Smith Orr (Eds.), *Improving leadership in nonprofit organizations* (pp. 234–251). San Francisco: Jossey-Bass.

Donaldson, S. I. (2005). Using program theory-driven evaluation science to crack the Da Vinci Code. *New Directions for Evaluation, 106*, 65–84.

Donaldson, S. I., & Bligh, M. C. (2006). Rewarding careers applying positive psychological science to improve quality of work life and organizational effectiveness. In S. I. Donaldson, D. E. Berger, & K. Pezdek (Eds.), *Applied psychology: New frontiers and rewarding careers* (pp. 277–295). Mahwah, NJ: Lawrence Erlbaum Associates.

Donaldson, S. I., & Christie, C. A. (2005). The 2004 Claremont Debate: Lipsey versus Scriven. Determining causality in program evaluation and applied research: Should experimental evidence be the gold standard? *Journal of Multidisciplinary Evaluation, 3*, 60–77.

Donaldson, S. I., & Christie, C. A. (2006). Emerging career opportunities in the transdiscipline of evaluation science. In S. I. Donaldson, D. E. Berger, & K. Pezdek (Eds.), *Applied psychology: New frontiers and rewarding careers* (pp. 243–259). Mahwah, NJ: Erlbaum.

Donaldson, S. I., Christie, C. A., & Mark, M. M. (in press). *What counts as credible evidence in evaluation and evidence-based practice?* Thousand Oaks, CA: Sage.

Donaldson, S. I., & Gooler, L. E. (2001). *Summary of the evaluation of The California Wellness Foundation's Work and Health Initiative.* Institute for Organizational and Program Evaluation Research, Claremont Graduate University, Claremant, CA.

Donaldson, S. I., & Gooler, L. E. (2002). Theory-driven evaluation of the work and health initiative: A focus on winning new jobs. *American Journal of Evaluation, 23*(3), 341–346.

Donaldson, S. I., & Gooler, L. E. (2003). Theory-driven evaluation in action: Lessons from a $20 million statewide work and health initiative. *Evaluation and Program Planning, 26*, 355–366.

Donaldson, S. I., Gooler, L. E., & Scriven, M. (2002). Strategies for managing evaluation anxiety: Toward a psychology of program evaluation. *American Journal of Evaluation, 23*(3), 261–273.

Donaldson, S. I., Gooler, L. E., & Weiss, R. (1998). Promoting health and well-being through work: Science and practice. In X. B. Arriaga & S. Oskamp (Eds.), *Addressing community problems: Research and intervention* (pp. 160–194). Newbury Park, CA: Sage.

Donaldson, S. I., Graham, J. W., & Hansen, W. B. (1994). Testing the generalizability of intervening mechanism theories: Understanding the effects of school-based substance use prevention interventions. *Journal of Behavioral Medicine, 17*, 195–216.

Donaldson, S. I., Graham, J. W., Piccinin, A. M., & Hansen, W. B. (1995). Resistance-skills training and onset of alcohol use: Evidence for beneficial and potentially harmful effects in public schools and in private Catholic schools. *Health Psychology, 14*, 291–300.

Donaldson, S. I., & Lipsey, M. W. (2006). Roles for theory in contemporary evaluation practice: Developing practical knowledge. In I. F. Shaw, J. C. Greene, & M. M. Mark (Eds.), *The Sage handbook of evaluation* (pp. 56–75). London: Sage.

Donaldson, S. I., & Scriven, M. (2003a). Diverse visions for evaluation in the new millennium: Should we integrate or embrace diversity? In S. I. Donaldson &

M. Scriven (Eds.), *Evaluating social programs and problems: Visions for the new millennium* (pp. 3–16). Mahwah, NJ: Lawrence Erlbaum Associates.

Donaldson, S. I. & Scriven, M. (Eds.). (2003b). *Evaluating social programs and problems: Visions for the new millennium.* Mahwah, NJ: Lawrence Erlbaum Associates.

Donaldson, S. I., Street, G., Sussman, S., & Tobler, N. (2001). Using meta-analyses to improve the design of interventions. In S. Sussman (Ed.), *Handbook of program development for health behavior research and practice* (pp. 449–466). Newbury Park, CA: Sage.

Donaldson, S. I., & Weiss, R. (1998). Health, well-being, and organizational effectiveness in the virtual workplace. In M. Igbaria & M. Tan (Eds.), *The virtual workplace* (pp. 24–44). Harrisburg, PA: Idea Group Publishing.

Dooley, D., Fielding, J., & Levi, L. (1996). Health and unemployment. *Annual Review of Public Health, 17,* 449–465.

Dooley, D., & Prause, J. (1998). Underemployment and alcohol abuse in the national longitudinal survey of youth. *Journal of Studies on Alcohol, 59,* 669–680.

Eisenberg, L., Winters, L., & Alkin, M. C. (2005). The case: Bunche-Da Vinci Learning Partnership Academy. In M. C. Alkin & C. A. Christie (Eds.), *Theorists' models in action: New directions for evaluation* (Vol. 106, pp. 5–13). San Francisco: Jossey-Bass.

Employment Development Department. (2000, September). Unemployment rates. Retrieved September 1, 2000, from htt://www.labormarketinfo.edd.ca.gov/cgi/dataanalysis/AreaSelection.asp?tableName=Labforce

Engle, M., & Altschuld, J. (2004). An update on university-based evaluation training. *The Evaluation Exchange, IX*(4), 9–10.

Fetterman, D. (2003). Empowerment evaluation strikes a responsive cord. In S. I. Donaldson & M. Scriven (Eds.), *Evaluating social programs and problems: Visions for the new millennium* (pp. 63–76). Mahwah, NJ: Lawrence Erlbaum Associates.

Fitzpatrick, J. L. (2002). Dialog with Stewart Donaldson. *American Journal of Evaluation, 3*(3), 347–365.

Fitzpatrick, J. L. (2004). Exemplars as case studies: Reflections on the links between theory, practice, and context. *American Journal of Evaluation, 25*(4), 541–559.

Fitzpatrick, J. L., Mark, M. M., & Christie, C. A. (in press). *Exemplars in program evaluation practice.* Thousand Oaks, CA: Sage.

Fitzpatrick, J. L., Sanders, J. R., & Worthen B. R. (2004). *Program evaluation: Alternative approaches and practical guidelines.* Boston, MA: Allyn & Bacon.

Freud, S. (1930). *Civilization and its discontents.* London: Hogarth.

Funnell, S. (1997). Program logic: An adaptable tool for designing and evaluating programs. *Evaluation News & Comment,* 5–17.

Gargani, J. (2003, November). *The history of theory-based evaluation: 1909 to 2003.* Paper presented at the American Evaluation Association annual conference, Reno, NV.

Ghere, G., King, J. A., Stevahn, L., & Minnema, J. (2006). A professional development unit for reflecting on program evaluator competencies. *American Journal of Evaluation, 27*(1), 108–123.

Greene, J. C. (2005). A valued-engaged approach to evaluating the Bunche–Da Vinci Learning Academy. In M. C. Alkin & C. A. Christie (Eds.), *Theorists' models in action* (pp. 27–46). San Francisco, CA: Jossey-Bass.

Guiding Principles for Evaluators (1994). *New Directions for Program Evaluation, 66.* San Francisco: Jossey-Bass.

Hansen, W. B. (1993). School-based alcohol prevention programs. *Alcohol Health & Research World, 17,* 54–60.

Hansen, W. B., & McNeal, Jr., R. B. (1996). The law of maximum expected potential effect: Constraints placed on program effectiveness by mediator relationships. *Health Education Research, 11,* 501–507.

Hasenfeld, Y. (1983). *Human service organizations.* Englewood Cliffs, NJ: Prentice-Hall.

Henry, G. T. (2005). In pursuit of social betterment: A proposal to evaluate the Da Vinci Learning Model. In M. C. Alkin & C. A. Christie (Eds.), *Theorists' models in action.* (pp. 47–64). San Francisco, CA: Jossey-Bass.

Henry, G. T., & Mark, M. M. (2003). Toward an agenda for research on evaluation. In C. A. Christie (Ed.), *The practice-theory relationship in evaluation* (pp. 69–80). San Francisco, CA: Jossey-Bass.

House, E. R. (2003). Stakeholder bias. In C. A. Christie (Ed.), *New directions for evaluation, No. 97* (pp. 53-56). San Franciso: Jossey-Bass.

Howard, A. (1995). A framework for work change (pp. 1–44). In A. Howard (Ed.), *The changing nature of work* (pp. 1–44). San Francisco, CA: Jossey-Bass.

Jin, R. L., Shah, C. P., & Svoboda, T. J. (1999). The impact of unemployment on health: A review of the evidence. *Journal of Public Health Policy, 18,* 275–301.

Joint Committee on Standards for Education Evaluation (1994). *The program evaluation standards: How to assess evaluations of educational programs.* Thousand, Oaks, CA: Sage.

Karasek, R., & Theorell, T. (1990). *Healthy work: Stress, productivity, and the reconstruction of working life.* New York: Basic Books.

King, J. A. (2005). A proposal to build evaluation capacity at the Bunch—Da Vinci Learning Partnership Academy. In M. C. Alkin & C. A. Christie (Eds.), *Theorists' models in action* (pp. 85–98). San Francisco, CA: Jossey-Bass.

Levin, H. M., & McEwan, P. J. (2001). *Cost-effectiveness analysis: Methods and applications.* Thousand Oaks, CA: Sage.

Levy, S. (March 2001). Five years of strong economic growth: The impact on poverty, inequality, and work arrangements in California. *The Future of Work and Health Program.* The California Wellness Foundation, San Francisco, CA.

Lincoln, Y. S. (2003). Fourth generation evaluation in the new millennium. In S. I. Donaldson & M. Scriven (Eds.), *Evaluating social programs and problems: Visions for the new millennium* (pp. 77–90). Mahwah, NJ: Lawrence Erlbaum Associates.

Lipsey, M. W. (1988). Practice and malpractice in evaluation research. *Evaluation Practice, 9,* 5–24.

Lipsey, M. W. (1990). *Design sensitivity.* Newbury Park, CA: Sage.

Lipsey, M. W. (1993). Theory as method: Small theories of treatments. *New Directions For Program Evaluation, 57,* 5–38.

Lipsey, M. W., & Cordray, D. (2000). Evaluation methods for social intervention. *Annual Review of Psychology, 51*, 345–375.

Lipsey, M. W., Crosse, S., Dunkel, J., Pollard, J., & Stobart, G. (1985). Evaluation: The state of the art and the sorry state of the science. *New Directions for Program Evaluation, 27*, 7–28.

Lipsey, M. W., & Pollard, J. A. (1989). Driving toward theory in program evaluation: More models to choose from. *Evaluation and Program Planning, 12*, 317–328.

Lipsey, M. W., & Wilson, D. B. (1993). The efficacy of psychological, educational, and behavioral treatment: Confirmation from meta-analysis. *American Psychologist, 48*, 1181–1209.

London, M., & Greller, M. M. (1991). Invited contribution: Demographic trends and vocational behavior: A twenty year retrospective and agenda for the 1990s. *Journal of Vocational Behavior, 28*, 125–164.

Mark, M. M. (2003). Toward an integrative view of the theory and practice of program and policy evaluation. In S. I. Donaldson & M. Scriven (Eds.), *Evaluating social programs and problems: Visions for the new millennium* (pp. 183–204). Mahwah, NJ: Lawrence Erlbaum Associates.

Mark, M. M., Henry, G. T., & Julnes, G. (2000). *Evaluation: An integrative framework for understanding guiding, and improving policies and programs.* San Francisco, CA: Jossey-Bass.

Marmot, M. G. (1998). Social differentials in health within and between populations. *Daedelus: Proceedings of the American Academy of Arts, 123*, 197–216.

McLaughlin, J. A., & Jordon, G. B. (1999). Logic models: A tool for telling your program's performance story. *Evaluation and Program Planning, 22*(1) 65–72.

Mercier, C., Piat, M., Peladeau, N., & Dagenais, C. (2000). An application of theory-driven evaluation to a drop-in youth center. *Evaluation Review, 24*, 73–91.

Mertens, D. M. (2003). The inclusive view of evaluation: Visions for the new millennium. In S. I. Donaldson & M. Scriven (Eds.), *Evaluating social programs and problems: Visions for the new millennium* (pp. 91–108). Mahwah, NJ: Lawrence Erlbaum Associates.

Mertens, D. M. (2005). The inauguration of the international organization for cooperation in evaluation. *American Journal of Evaluation, 26*(1), 124–130.

Mouton, J., & Wildschut, L. (2006). Theory driven realist evaluation. In R. Basson, C. Potter, & M. Mark (Eds.), *Internationalizing evaluation: Reflections on evaluation approaches and their use globally using South Africa as a case example.* Manuscript in preparation.

Patton, M. Q. (1997). *Utilization-focused evaluation: The new century text* (3rd ed.).Thousand Oaks, CA: Sage.

Patton, M. Q. (2001). Evaluation, knowledge management, best practices, and high quality lessons learned. *American Journal of Evaluation, 22*, 329–336.

Pawson, R. & Tilley, N. (1997). *Realist evaluation.* Thousand Oaks, CA: Sage.

Pelletier, K. (1993). A review and analysis of the health and cost-effective outcome studies of comprehensive health promotion and disease prevention programs at the worksite, 1991–1995 update. *American Journal of Health Promotion, 8*, 380–388.

Quick, J., Murphy, L. R., Hurrell, J., & Orman, D. (1992). The value of work, the risk of distress, and the power of prevention. In J. Quick, L. Murphy, & J. Hurrell, Jr. (Eds.), *Stress and well-being at work.* Washington, DC: American Psychological Association.

Reichhart, C., & Rallis, C. S. (Eds.). (1994). The qualitative–quantitative debate: New perspectives. *New Directions for Program Evaluation, p. 61.*

Renger, R., & Titcomb, A. (2002). A three-step approach to teaching logic models. *American Journal of Evaluation, 23(4),* 493–503.

Reynolds, A. J. (1998). Confirmatory program evaluation: A method for strengthening causal inference. *American Journal of Evaluation, 19,* 203–221.

Rossi, P. H. (2004). My views of evaluation and their origins. In M. C. Alkin (Ed.), *Evaluation roots* (pp. 122–131). Thousand Oaks, CA: Sage.

Rossi, P. H., Lipsey, M. W., & Freeman, H. E. (2004). *Evaluation: A systematic approach* (7th ed.). Thousand Oaks, CA: Sage.

Rousseau, D. M. (1996). Changing the deal while keeping people. *Academy of Management Executive, 10,* 50–59.

Russon, C. (2004). Cross-cutting issues in international standards development. *New Directions for Evaluation, 104,* 89–93.

Sanders, J. R. (2001). A vision for evaluation. *American Journal of Evaluation, 22,* 363–366.

Schauffler, H., & Brown, R. (2000). *The state of health insurance in California, 1999.* Berkeley, CA: Regents of University of California.

Scriven, M. (1991). *Evaluation thesaurus* (4th ed.). Thousand Oaks, CA: Sage.

Scriven, M. (1998). Minimalist theory: The least theory that practice requires. *American Journal of Evaluation, 19,* 57–70.

Scriven, M. (2003). Evaluation in the new millennium: The transdisciplinary vision. In S. I. Donaldson & M. Scriven (Eds.), *Evaluating social programs and problems: Visions for the new millennium* (pp. 19–42). Mahwah, NJ: Lawrence Erlbaum Associates.

Seigart, D., & Brisolara, S. (2002). *Feminist evaluation: Explorations and experiences.* San Francisco, CA: Jossey-Bass.

Shadish, W. R. (1993). Critical multiplism: A research strategy and its attendant tactics. *New Directions for Program Evaluation, 60,* 13–57.

Shadish, W. R. (1998). Presidential address: Evaluation theory is who we are. *American Journal of Evaluation, 19*(1), 1–19.

Shadish, W. R., Cook, T. D., & Campbell, D. T. (2002). *Experimental and quasi-experimental designs for generalized causal inference.* Boston: Houghton-Mifflin.

Shadish, W. R., Cook, T. D., & Leviton, L. C. (1991). *Foundations of program evaluation: Theories of practice.* Newbury Park, CA: Sage.

Slater, J. K. (2006). Book review: Evaluating social programs and problems: Visions for the new millennium. *American Journal of Evaluation, 27*(1), 128–129.

Stame, N. (2004). Theory-based evaluation and types of complexity. *Evaluation: The International Journal of Theory, Research & Practice, 10*(1), 58–76.

Stern, E. (2004). What shapes European evaluation? A personal reflection. *Evaluation: The International Journal of Theory, Research & Practice, 10(1),* 7–15.

Stufflebeam, D. L. (Ed.). (2001). Evaluation models. *New Directions for Evaluation, 89.* San Francisco, Jossey-Bass.

Thompson-Robinson, M., Hopson, R., & SenGupta, S. (Eds.). (2004). *In search of cultural competence in evaluation: Toward principles and practice.* San Francisco, CA: Jossey-Bass.

Tilley, N. (2004). Applying theory-driven evaluation to the British Crime Reduction Programme: The theories of the programme and of its evaluations. *Criminal Justice: International Journal of Policy & Practice, 4(3),* 255–276.

Triolli, T. (2004). Review of "Evaluating social program and problems: Visions for the new millennium." *Evaluation and Program Planning, 26(3),* 82–88.

U.S. Department of Health and Human Services. (1993). 1992 national survey of worksite health promotion activities: Summary. *American Journal of Health Promotion, 7,* 452-464.

van der Knaap, P. (2004). Theory-based evaluation and learning: Possibilities and challenges. *Evaluation: The International Journal of Theory, Research & Practice, 10(1),* 16–34.

Vinokur, A. D., van Ryn, M., Gramlich, E., & Price, R. H. (1991). Long-term follow-up and benefit-cost analysis of the JOBS program: A preventive intervention for the unemployed. *Journal of Applied Psychology, 76,* 213–219.

Weiss, C. H. (1997). How can theory-based evaluation make greater headway? *Evaluation Review, 21,* 501–524.

Weiss, C. H. (1998). *Evaluation: Methods for studying programs and policies* (2nd ed.). Upper Saddle River, NJ: Prentice Hall.

Weiss, C. H. (2004a). On theory-based evaluation: Winning friends and influencing people. *The Evaluation Exchange, IX, 4,* 1–5.

Weiss, C. H. (2004b). Rooting for evaluation: A cliff notes version of my work. In M. C. Alkin (Ed.), *Evaluation roots* (pp. 153–168). Thousand Oaks, CA: Sage.

West, S. G., & Aiken, L. S. (1997). Toward understanding individual effects in multi-component prevention programs: Design and analysis strategies. In K. J. Bryant, M. Windle, & S. G. West (Eds.), *The science of prevention: Methodological advances from alcohol and substance abuse research* (pp. 167–209). Washington, DC: American Psychological Association.

Wholey, J. S. (2003). Improving performance and accountability: Responding to emerging management challenges. In S. I. Donaldson & M. Scriven (Eds.), *Evaluating social programs and problems: Visions for the new millennium* (pp. 43–62). Mahwah, NJ: Lawrence Erlbaum Associates.

AUTHOR INDEX

A

Adkins, J. A., 53
Adler, N. E., 52
Aiken, L. S., 28, 235
Aley, J., 54
Alkin, M. C., 4, 18, 175, 176, 190,
  233, 234, 239, 241, 244
Altschuld, J., 244

B

Bamberg, S., 235
Baron, R. M., 30
Bickman, L., 14, 22, 235
Bligh, M. C., 52, 194
Brisolara, S., 4
Brousseau, R., 52, 53, 54
Brown, R., 53

C

Campbell, B., 209
Campbell, D. T., 4, 8, 14, 200, 202, 235
Chen, H. T., 4, 6, 8, 11, 14, 21, 27, 41,
  44, 57, 186, 199, 235
Christie, C. A., 4, 8, 18, 21, 175, 190,
  199, 209, 210, 232, 233, 234, 238,
  239, 241, 242, 244
Cook, T. D., 4, 5, 6, 7, 8, 14, 57, 200,
  202, 235, 245
Cordray, D., 14
Cousins, J. B., 209
Crano, W. D., 8
Crosse, S., 27
Cummings, T. G., 16

D

Dagenais, C., 235
Davidson, E. J., 44
Donaldson, S. I., 4, 7, 8, 9, 10, 11,
  14, 15, 16, 17, 18, 21, 22, 23,
  26, 27, 32, 35, 44, 45, 51, 52,
  53, 54, 57, 61, 82, 186, 188,
  190, 194, 195, 196, 199, 209,
  210, 212, 223, 224, 225, 226,
  227, 228, 229, 232, 235, 238,
  239, 241, 242, 244
Dooley, D., 52, 53
Dunkel, J., 27

E

Eisenberg, L., 176
Engle, M., 244

F

Fetterman, D., 4, 8
Fielding, J., 52, 53
Fitzpatrick, J. L., 14, 32, 41, 44,
  186, 195, 210, 224,
  225, 244
Freeman, H. E., 4, 7, 21, 23, 41, 44,
  186, 190, 197
Freud, S., 51
Funnell, S., 24, 26

G

Gargani, J., 9
Ghere, G., 244

SUBJECT INDEX

About The Author

Stewart I. Donaldson is Professor and Chair of Psychology, Director of the Institute of Organizational and Program Evaluation Research, and Dean of the School of Behavioral and Organizational Sciences at Claremont Graduate University. Dr. Donaldson continues to develop and lead one of the most extensive and rigorous graduate programs specializing in applied psychological and evaluation science. Since the mid-1990s, he has taught numerous university courses, professional development workshops, and mentored and coached more than 100 graduate students and working professionals. Dr. Donaldson has also provided organizational consulting, applied research, and program evaluation services to more than 100 different organizations. He has been the principal investigator on more than 20 extramural grants and contracts totaling approximately $4 million to support research, evaluations, scholarship, and graduate students at Claremont Graduate University. Dr. Donaldson serves on the editorial boards of the *American Journal of Evaluation*, *New Directions for Evaluation*, and the *Journal of Multidisciplinary Evaluation*, cofounded and leads the Southern California Evaluation Association, and has served as cochair of the Theory-Driven Evaluation and Program Theory Topical Interest Group of the American Evaluation Association for eight years. He has authored and coauthored more than 200 evaluation reports, scientific journal articles, and book chapters. Recent books include: *Evaluating Social Programs and Problems: Visions for the New Millennium* (Lawrence Erlbaum Associates, 2003, with Michael Scriven), *Applied Psychology: New Frontiers and Rewarding Careers* (Lawrence Erlbaum Associates, 2006, with Dale E. Berger and Kathy Pezdek), *Program Theory-Driven Evaluation Science: Strategies and Applications* (Lawrence Erlbaum Associates, 2007, this volume). Dr. Donaldson has been honored with Early Career Achievement Awards from the Western Psychological Association and the American Evaluation Association.